Tributes
Volume 26

Learning and Inferring
Festschrift for Alejandro C. Frery
on the Occasion of his 55th Birthday

Tributes Series Editor
Dov Gabbay   dov.gabbay@kcl.ac.uk

Learning and Inferring

Festschrift for Alejandro C. Frery
on the Occasion of his 55th Birthday

edited by

Bruno Lopes

and

Talita Perciano

ISBN 978-1-84890-171-1

College Publications
Scientific Director: Dov Gabbay
Managing Director: Jane Spurr

http://www.collegepublications.co.uk

Cover design by Laraine Welch

Printed by Lightning Source, Milton Keynes, UK

Contents

Preface

In 1993, a Ph.D. Thesis presented the development of computational tools for image simulation, processing, and analysis of synthetic aperture (SAR) radar images. These tools were proposed based on statistical modelling inside a bayesian context. That was the beginning of the academic career of Prof. Alejandro C. Frery, one of the most prominent statisticians of today. He achieved this status not only by the technical accuracy and quality of his work but also by the kindness of sharing his knowledge and by all of his efforts aiming the progress of science.

Learning and Inferring is an abstract reference to some of the many areas wherein Prof. Frery applied efforts to evolve the human knowledge. The idea for this book came from the desire to celebrate the contributions of Prof. Frery's work in probability and statistics. This briliant academic researcher is descendent of some of the most recognized scientists of the world, like Simeon Denis Poisson, Joseph Louis Lagrange, Pierre-Simon Laplace, Leonard Euler, Johann Bernoulli, Gottfried Wilhelm Leibniz, Nicolaus Copernicus and many other great names of science. We have selected nine papers associated with Prof. Frery's work, in an extended sense of this volume as we describe in here.

Loureiro and Ramos discuss contributions related to wireless sensor networks and show how Prof. Frery's work goes into this direction. Lima et al. present how statistical techniques, including platforms evaluated by Prof. Frery, may lead to data achieved from clustering techniques of colorectal cancer patients. This work aims to propose inferences on prognostic analysis of colorectal cancer. Cribari-Neto and Lima analyze a test for beta regression model, which is useful for modeling responses, such as rates and proportions. Flesia et al. revise several methods for the simulation of correlated clutter with desirable marginal law and correlation structure. Regarding the quantification of uncertainty, Lopes and Barbosa evaluate some metrics to quantify the concentration of probability density functions curve, discussing about how metrics are adequate to some different scenarios. In the context of polarimetric radar image processing, Torres et al. show how the use of stochastic distances derived from an appropriate statistical modeling, and its associated statistical tests, is highly relevant and promising. In the same context, Horta et al. explore the research on semi-supervised learning for PolSAR image classification. The performance analysis of three statistical clustering techniques adapted to a semi-supervised learning is presented. The problem of edge detection in PolSAR images is investigated by Sandri et al., where the recently proposed gravitational edge detection method is modified using non-standard neighbourhood configurations improving the obtained results. Finally, Siqueira et al. present several new strategies for a better surface data visualization. The proposed strategies are illustrated for different applications, showing its potential use in real life cases.

The goal of this book is not only honor Prof. Frery but also show how far he has been in science by means of his own contributions and by the school that has been rising around him. During this process we had to plunge into many fields, and the greatest challenge was to represent a wide amount of domains in which Alejandro is related. We know that this task was not fully completed (maybe it is not even possible in only one book!). However, we assure the contributions in here, kindly assigned by the authors, honor his name and represent a huge part of the kernel of Alejandro's way of doing science.

Teaching and educating early career researchers is on top of the most important contributions of Prof. Frery. He advised almost a hundred students by now, leading to a valuable contribution to the future of science. The two editors of this volume are, proudly, included into this select group, aiming to proceed with this long-term and (hopefully) neverending lineage of researchers. Along with collaborators, former students and friends, we offer this book as a token of our esteem and appreciation, and as a powerful wish that Alejandro continues to offer us his valuable contributions and knowledge for yet a long time.

Bruno Lopes (UFF, Brazil)
Talita Perciano (LBNL, USA)

Acknowledgements

We would like to thank the authors who joined us in this book and Enilson Costa, for the front-cover photography of Prof. Alejandto C. Frery. We would like also to thank the agencies/companies that supported the works presented in this book: CNPq, CAPES, FINEP, FAPEAL, FAPEMIG, FAPESP, SeCyT-UNC, CONICET, Petrobrás and Microsoft.

Thank you all for the contributions to this unique project!

Advances in the design of
wireless sensor networks[‡]

Heitor S. Ramos[*]   Antonio A. F. Loureiro[†]

[*] Instituto de Computação
Universidade Federal de Alagoas – UFAL
heitor@ic.ufal.br

[†] Departamento de Ciência da Computação
Universidade Federal de Minas Gerais – UFMG
loureiro@dcc.ufmg.br

1 Introduction

Wireless communication has impacted our society in a profound way. Very few advances in technology have been able to *shrink* the world in such a way as we have testified since the beginning of the last century. We live in a time where there is a plethora of wireless technologies that are the substrata of different wireless networks, which in turn are the fundamental components of pervasive and ubiquitous computing.

An important wireless network is the wireless sensor network (WSN), which is responsible for collecting data from the environment and sending it to a reporting site where data can be observed and analyzed. A sensor network is a basic building block of pervasive and ubiquitous computing. WSNs continue to expand and increase in both practical and research domains, due to new developments in hardware, software, and communication technologies. Nowadays, these networks are being used to gather information in real-life applications such as intelligent buildings, infrastructure (e.g., bridges, roads, grids), agriculture, security, among others.

Usually, one of the first problems to be solved in any network is the data communication problem, i.e., how two or more network entities communicate among themselves. In the case of a wireless network, in general, and of a wireless sensor network, in particular, we need to consider the network topology to solve the data communication problem.

The network topology represents the organization of the different network elements (e.g., nodes and links). It denotes the physical and logical

[‡]The authors thank CNPq, FAPEAL and FAPEMIG for research grants.

topological structure of a network. The physical topology depicts the placement of the various network elements, while the logical topology describes the data flows within a network, regardless of its physical design. Given a network topology, some of the factors that influence the data communication in a WSN are distances between nodes, available energy of nodes, physical interconnections and transmission rates.

The lifetime of a wireless sensor network depends on the network topology in different ways. It depends on the available sensors at the first hop from the sink node, leading to the "energy hole" problem (see Section 2.1). This happens because of the relay task that is more concentrated on nodes that are placed close to the sink node, when data collection algorithms are used.

A deeper understanding of this problem shows that, for a given network topology, a suitable centrality metric may be able to characterize the energy hole problem (see Section 2.2). This means that it is possible to determine, beforehand, which node is most likely to drain its battery due to the relay task.

This fact becomes very useful if we can take it to the operation space, i.e., if we can calculate a suitable centrality metric in a distributed way, given the disadvantages of designing a centralized solution for a wireless sensor network (see Section 2.3). This distributed solution may be used by different algorithms and protocols that can take advantage of the centrality metric. As we will see, this metric presents a much better representation of the traffic pattern in a wireless sensor network than other metrics such as betweenness.

As in any other domain, research has the potential to propose new developments, refinements and continuous improvements of current solutions to push the technology even further. Frery's contributions towards those three problems mentioned above (topology design, centrality metric and associate distributed algorithm) have advanced the state of the art and helped establishing new frontiers in the design of wireless sensor networks. It has been a great pleasure for us to collaborate with him.

In the following sections, we briefly describe Frery's work in those areas.

2 Contributions on stochastic modeling and analysis of wireless sensor networks

2.1 Stochastic topology model for wireless sensor networks

A Wireless sensor Network (WSN) is an ad hoc wireless network consisting of spatially distributed autonomous devices that cooperatively monitor environmental conditions such as temperature, pressure, and pollutants. WSNs have been studied in various application areas, e.g., health, military, home (Akyildiz et al., 2002; Culler et al., 2004), where human presence is either impossible or inadequate (Cui et al., 2006; Younis et al., 2006).

A WSN is a special case of a wireless ad hoc network where nodes are able to communicate when they are in the range of each other's wireless communication channel. This communication behavior can be represented by a geographical threshold graph, which means that nodes are distributed in a Euclidean space, and edges are assigned according to a threshold function involving the distance between nodes. In a WSN, this threshold is determined by the node's communication radii.

Wu et al. (2008) show that the lifetime of a uniformly deployed WSN is strongly limited by the sensors at the first hop from the sink, a problem known as "energy hole". This problem follows from the relay task that is more concentrated on nodes that are placed close to the sink node, when data collection algorithms are used.

Most WSN models in the literature assume that the network is comprised of homogeneous nodes, i.e., all sensors have the same capabilities in terms of energy, processing, memory, and communication. However, Yarvis et al. (2005) show that homogeneous ad hoc networks suffer from fundamental limitations and, hence, exhibit poor network performance such as end-to-end success rate, latency and energy consumption. Another class of WSN models assumes that there are different sets of nodes, each one with different capabilities. For instance, suppose we have two sets of nodes: the first one comprised of a small number of powerful high-end sensors (H-sensors), and the second one of a large number of low-end sensors (L-sensors). In this case, we have a Heterogeneous Sensor Network model (Yarvis et al., 2005).

The energy hole problem is also present in heterogeneous networks. In this case, it appears in the neighborhood of each H-sensor and the sink. The network lifetime cannot be desirably prolonged by just randomly increasing the number of sensors when a totally random deployment is used (Wu et al., 2008). The entire network lifetime can be improved by spreading more nodes nearby the sink in the homogeneous network, and around the H-sensors, in the heterogeneous case.

Point processes are stochastic models that describe the location of points in space. They are useful in a broad variety of scientific applications as, for instance, ecology, medicine and engineering (Baddeley, 2007). This modeling tool is useful to describe different types of deployment in WSNs. For instance, the isotropic stationary Poisson model, also known as fully random or uniformly distributed, is the basic point process mostly and is the mostly used modeling tool in WSNs.

The Poisson point process over a finite region $W \subset \mathbb{R}^2$ is defined by the following properties:

1. The probability of observing $n \in \mathbb{N}_0$ points in any set $A \subset W$ follows a Poisson distribution: $\Pr(N_A = n) = e^{-\eta\mu(A)}[\eta\mu(A)]^n/n!$, where $\eta > 0$ is the intensity and $\mu(A)$ is the area of A.

2. Random variables describing the number of points in disjoint subsets are independent.

In order to draw a sample from a Poisson point process with intensity $\eta > 0$ on a, without loss of generality, squared window $W = [0, \ell] \times [0, \ell]$, first sample from a Poisson random variable with mean $\eta\ell^2$; assume n was observed. Now obtain $2n$ samples from independent identically distributed random variables with uniform distribution on $[0, \ell]$, say $x_1, \ldots, x_n, y_1, \ldots, y_n$. The n points placed at coordinates $(x_i, y_i)_{1 \leq i \leq n}$ are an outcome of the Poisson point process on W with intensity η. If n is known beforehand, rather than the outcome of a Poisson random variable, then the n points placed at coordinates $(x_i, y_i)_{1 \leq i \leq n}$ are an outcome of the Binomial point process on W; this last process is denoted $B(n)$.

The Matérn's Simple Sequential Inhibition process can be defined iteratively as the procedure that tries to place n points in W. The first point is placed uniformly, and until all the n points are placed or the maximum number of iterations $t_{\max}$ is reached, a new location is chosen uniformly on W. A new point is placed there if the new location is not closer than r to any previous point; otherwise the location is discarded, the iteration counter is increased by one and a new location is chosen uniformly. At the end, there are $m \leq n$ points in W that lie at least r units from each other. This process describes the distribution of non-overlapping discs of radii $r/2$ on W; denote it $M(n, r)$.

First Frery's contributions towards a stochastic node deployment modeling of WSNs was introduced in Frery et al. (2008) and Frery et al. (2010), where the $\mathcal{C}$ model was proposed based on an inhomogeneous Poisson point process suitable for homogeneous networks. $\mathcal{C}$ model is an attractive

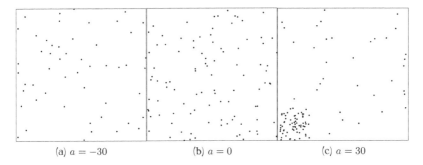

(a) $a = -30$ (b) $a = 0$ (c) $a = 30$

Figure 1: Samples of 100 spatial point processes

process built by merging two Poisson processes with different intensities. A step point process in $W' \subset W \subset \mathbb{R}^2$ with parameters $a, \lambda > 0$ is defined as two independent Point processes: one with parameter λ on $W \setminus W'$, and other with parameter $a\lambda$ on W'. Denote this process $S(n, a)$.

Without loss of generality, it was defined the compound point process on $W = [0, 100]^2$, $W' = [0, 25]^2$ and $\eta = 1$, denoted by $\mathcal{C}(n, a)$, as

$$
\mathcal{C}(n, a) = \begin{cases} M(n, r_{\max}(1 - e^a)) & \text{if } a < 0 \\ B(n) & \text{if } 0 \leq a \leq 1 \\ S(n, a) & \text{if } a > 1. \end{cases}
$$

where $r_{\max}$ is the maximum exclusion distance set to $r_{\max} = n^{-1/2} = 1/10$. The $\mathcal{C}(n, a)$ point process spans in a seamless manner the repulsive ($a < 0$, Figure 1a), full random ($a \in [0, 1]$, Figure 1b) and attractive cases ($a > 0$, Figure 1c). For the sake of completeness $\mathcal{C}(n, -\infty)$ denotes the deterministic placement of n regularly spaced sensors on W at the maximum possible distance among them. Samples from the $\mathcal{C}$ process can be easily generated using basic functions from the spatstat package for R (Baddeley & Turner, 2005).

Repulsive processes are able to describe the intentional though not completely controlled location of sensors as, for instance, when they are deployed by a helicopter at low altitude. Sensors located by a binomial process could have been deployed from high altitude, so their location is completely random and independent of each other. Attractive situations may arise in practice when sensors cannot either be deployed or function everywhere as, for instance, when they are spread in a swamp: those that fall in a dry spot survive, but if they land on water they fail to function.

Frery's work show that $\mathcal{C}$ is able to represent a wide variety of WSNs deployments, for instance, when $a > 0$, we observe that the nodes will present higher density around the sink, helping on alleviate the energy hole problem.

The M^2P^2 model was firstly introduced in Ramos et al. (2011) to evolve the C model making a more general one, and take into account heterogeneous networks that present two tier of nodes, namely, L- and H-sensors. M^2P^2 is a node deployment model that represents a wide variety of of topologies that may exhibit desirable characteristics for WSNs. It was shown that a proper planned stochastic deployment leads to improve the network performance by means of shorter average path length and higher cluster coefficient (which can improve the fault-tolerance properties). Besides, an appropriated deployment can properly address the energy hole problem.

This process starts firstly placing m H-sensors on W, and then deploying the remaining $n - m$ sensors "close" to them. Denote the coordinates of the m H-sensors by $\boldsymbol{h} = \{(hx_1, hy_1), \ldots, (hx_m, hy_m)\}$ (how these sensors are placed will be described later).

Without loss of generality, in the following, we consider the intensity function λ, which has increased but constant intensity around selected spots:

$$\lambda(x, y) = \begin{cases} a, & \text{if } d((x, y), (hx_i, hy_i)) \leq r_c, 1 \leq i \leq m, \\ 1, & \text{otherwise.} \end{cases} \tag{1}$$

where $a \geq 1$ (the attractiveness parameter), d is any distance measure, and r_c is the communication radius of the low-end sensors (L-sensors). In the remaining of this work, we employ the Euclidean distance but any suitable distance measure may be used to enhance realism. Denote such process by $\Lambda(n - m, a, \boldsymbol{h})$.

Notice that a stochastic point process defined by an intensity function λ as the one in Equation (1) has overall mean intensity given by $\int_W \lambda$. If $a > 1$, then it is more likely to have points around the m coordinates where there is an H-sensor; if A_1 belongs to the area of influence of an H-sensor and A_2 does not, but $\mu(A_1) = \mu(A_2)$, on average there will be a more sensors in the former than in the latter subset. As defined, two or more H-sensors that are arbitrarily close will behave as a single H-sensor for the deployment of L-sensors, since the Λ process favors the occurrence of the latter as a function of the distance to the former.

Note that if $\lambda(x, y) = \lambda$, the inhomogeneous Poisson point process becomes the basic Poisson point process, i.e., it reduces to complete randomness.

In heterogeneous WSNs, H-sensors are useful to provide long-range shortcuts and diminish the number of hops to reach the sink node. They have a high-powerful radio that is able to communicate in long-range distances and a high-capacity battery that increases their lifetime. Those features make the H-sensors more expensive than the other nodes. It is desirable that the deployment of the H-sensors be made in such way that it diminishes the amount of H-sensors close to each other and, thus, decreases the total amount of H-sensors required to create the appropriate shortcuts.

The SSI (Simple Sequential Inhibition) stochastic point process (Baddeley, 2007) is a convenient model for the repulsive deployment of sensors. This process is defined on a window W by the maximum number of m points and an inhibition distance d. The first of the m points is placed in W obeying a binomial process. At each subsequent iteration, a new point is placed in W and it is accepted only if all other previous points lie further than d, otherwise it is rejected. The procedure stops either when the m points have been placed or when a maximum number of iterations is reached. This is the process that places at most m non-overlapping disks of radii $d/2$ on W. There are richer repulsive point processes, where there is no strict inhibition as, for instance, the Strauss process (Baddeley, 2007); the SSI suffices and is denoted as $H(m, 2r)$.

Thus, the M^2P^2 model can be defined as:

Definition 2.1 ($M^2P^2(m, n, a, r_c, r_i)$ on $W \subset \mathbb{R}^2$**)** *Consider a number $m \geq 1$ of H-sensors over a total of $n > m$ sensors, the intensity $a \geq 1$ of L-sensors on a circle or radius $r_c > 0$ centered at each H-sensor (r_c is the communication radius among L-sensors) and inhibition radius $r_i > 0$ among H-sensors. Thus, M^2P^2 is a compounded process of m samples of $H(m, 2r_i)$ (the H-sensors) and $n - m$ samples of $\Lambda(n - m, a, \boldsymbol{h})$ (the L-sensors), $\boldsymbol{h}$ is the set of m coordinates of the H-sensors.*

Firstly, take a sample from an $H(m, r_i)$ process with exactly m points: the coordinates of the m H-sensors. Secondly, return the outcome of an inhomogeneous binomial point process through the intensity function λ defined in Equation (1) using as $\boldsymbol{h}$ the m coordinates obtained in the first step, and take a sample of $n - m$ points by using $\Lambda(n - m, a, \boldsymbol{h})$.

Figure 2 shows four outcomes of the M^2P^2 process with 300 nodes inside an area of 100×100 square units: 1 (first), 10 (second and third), and 15 (forth) H-sensors (dark points) with attractively deployed L-sensors around the H-sensors with attractiveness parameters 5 (first), and 15 (third and forty). The leftmost outcome shows an homogeneous network ($m = 1$) where the darker point represents the sink node. The other figures show three heterogeneous WSNs. If $a = 1$, no attractive behavior around the

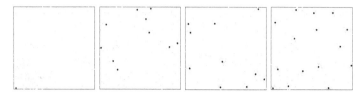

Figure 2: Outcomes of M^2P^2 for 300 nodes with 1, 10, 10 and 15 H-sensors (in black) and attractiveness 15, 5, 10 and 15, respectively

sink and the H-sensors is taken.

The M^2P^2 process can be extended in two ways, namely the deployment of the H-sensors and the deployment of the simple nodes (L-sensors). It is also immediate to generalize it to higher dimensions.

A sample from the M^2P^2 process is just a set of marked points. The connectivity radii among L- and H-sensors, r_c and r_{ch} respectively,induce a network topology.

Figure 3 shows two outcomes of the geographical threshold graphs that represents the network induced by the M^2P^2 model. Black points represent H-sensors, gray points represent L-sensors, the triangle represents the sink node. Edge colors follow the node type. There are 1000 sensors, being 30 H-sensors and 970 L-sensors deployed in a 1000×1000 sensor field. The communication radii are $r_c = 50$ and $r_{ch} = 300$ for L- and H-sensors, respectively, and $a = 5$. Observe that heterogeneous WSNs like the ones present in Figure 3 can be seen as an union of homogeneous WSNs, where the H-sensors act like sinks and are able to communicate in a different wireless channel acting as an overlay network (or even can be extended to communicate using another technology such as Internet, cable, etc.). Those kind of networks can also be seen as a multi-sink approach, considering the sink a more powerful element. This model is more likely to be used in large scale WSNs as it scales like a hierarchical WSNs. In such networks, the energy hole problem happens in the L-sensors. We are disregarding the energy hole in the H-sensors as we assume they present a powerful battery or they can be plugged to a continuous energy source.

In this field, Frery's C model, and other studies that extended this model, are able to represent a wide variety of WSNs topologies from totally random to planned stochastic node deployment in both heterogeneous and homogeneous sensor networks. Those models can be used in network simulators to represent more realistic scenarios than the usual uniformly random deployment. They are also useful to guide the network designer to chose the most suitable topology for the desirable application. It can

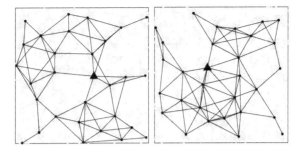

Figure 3: Two outcomes of network graphs generated by the M^2P^2 model. Dark points are the H-sensors, gray points are the L-sensors and the triangle is the sink node

be used to beforehand assess the energy hole behavior. For instance, for a given scenario, it was shown by simulating the M^2P^2 model that around only 3% of H-sensors (50 of 1500) and deploying nodes by using the slightly attractive L-sensors around the repulsive H-sensors model suffice to achieve high coverage and to alleviate the energy hole problem. An homogeneous uniformly random deployment achieves similar coverage with only 1000 node but fails to tackle the energy hole problem.

2.2 Characterization of wireless sensor networks

During the studies on deployment models as described in Section 2.1, it was observed that, for a given network, a suitable centrality metric may be able to characterize the energy hole problem. Thus, it is possible to determine, beforehand, which node is most likely to drain its battery due to the relay task.

There are many metrics proposed in the literature for characterizing and representing complex networks (see, for instance, Costa et al., 2007). Consider a network whose topology is represented by the geographical threshold graph $G(\boldsymbol{V}, \boldsymbol{E})$, where $\boldsymbol{V} = \{v_1, \ldots, v_n\}$ is the set of $|\boldsymbol{V}| = n$ nodes, and $\boldsymbol{E}$ is the set of edges.

Depending on the underlying communication model, WSNs can be represented by directed or undirected graphs. Thus, let us define the in- and out-neighborhoods of node v_i as $\boldsymbol{N}_i^{\text{in}} = \{v_j : e_{ji} \in \boldsymbol{E}\}$ and $\boldsymbol{N}_i^{\text{out}} = \{v_j : e_{ij} \in \boldsymbol{E}\}$, respectively. The neighborhood of a vertex v_i is $\boldsymbol{N}_i = \boldsymbol{N}_i^{\text{in}} \cup \boldsymbol{N}_i^{\text{out}}$. The in- and out-degrees of a vertex are defined as $k_i^{\text{in}} = |\boldsymbol{N}_i^{\text{in}}|$ and $k_i^{\text{out}} = |\boldsymbol{N}_i^{\text{out}}|$, respectively. The degree of a vertex v_i is defined as $k_i = |\boldsymbol{N}_i|$. Edges may be weighted, i.e., there may be a function $W : E \to \mathbb{R}$ which associates a real-valued weight w_e to every $e \in E$.

In the WSN context, it is useful to find strong relationships among topological metrics from the complex network theory and network metrics. For instance, Ramos et al. (2014) study the relationship between energy consumption and topological metrics that define the centrality of a node in a specific context.

There are several indices of centrality based on different graph features such as distance between vertices, *Closeness* (Beauchamp, 1965; Sabidussi, 1966), degree, *Eccentricity* (Hage & Harary, 1995), neighborhood importance, *Eigenvector* (Bonacich, 1972), *Hub Score, Authority* (Kleinberg, 1999) and *Page Rank* (Brin & Page, 1998). Another widely used concept in indices of centrality is the graph shortest path; for example, the *Shortest-path Betweeness Centrality* (Freeman, 1977, 1978–1979) calculates the centrality of vertex i based on the proportion of the number of geodesics (shortest paths) between any pair of vertices that falls on i by the total number of geodesics in the graph.

Locating and counting geodesics is difficult with large networks (Freeman, 1978–1979), and computational resources are limited in WSNs. The most efficient centralized algorithm to calculate Betweenness has running time $O\left(nm + n^2 \log n\right)$ for weighted graphs, and $O\left(nm\right)$ for unweighted graphs, where n and m are the number of vertices and edges respectively. The Betweenness of node v is defined as:

$$\mathrm{B}(v) = \sum_{s=1}^{n} \sum_{t=1}^{n} \frac{\sigma_{st}(v)}{\sigma_{st}}, \tag{2}$$

where σ_{st} is the number of shortest paths from s to t, $\{s,t\} \in \mathbf{V}$, and $\sigma_{st}(v)$ is the number of shortest paths from s to t that pass through $v \in \mathbf{V}$, $s \neq v \neq t$ and $s \neq t$.

In WSN scenarios, communication typically takes place between sensor nodes and the sink node, and vice versa. In order to consider this characteristic, we adopt a new centrality metric, namely Sink Betweenness (SBet – Oliveira et al., 2010; Ramos et al., 2012), which considers only the shortest paths that include the sink as one of the terminal nodes. It is defined, for every $v \in V$, as

$$\mathrm{SBet}(t) = \sum_{i \in \psi_t} \frac{\sigma_{ts}}{\sigma_{is}}, \tag{3}$$

where s is the sink, σ_{ts} is the number of shortest paths from t to the sink, σ_{is} is the number of shortest paths from i to the sink, $\psi_t = \{i \in V | t \in SP_{i \to s}\}$, and $SP_{i \to s}$ is the set of all shortest-paths from a node i to the sink, so ψ_t is the set of nodes that contains t at least in one of their shortest-paths.

When considering a scenario where nodes communicate to the sink by (data collection) by using a shortest path based routing algorithm, only betweenness and sink betweenness were able to capture the energy hole behavior. For instance, Figure 4 shows the correlograms for 100 and 400 nodes, with the sink both centered and randomly placed. Notice that Sink Betweenness presents higher correlation values than Betweenness and is less sensitive to the sink's location than the Betweenness.

In this context, the SBet metric showed to be useful in the design space of WSNs where deployment models can be assessed by means of the SBet metric to chose the most suitable topology that diminishes the energy hole problem. It was observed that if the SBet distribution for a given network resembles a uniform random distribution, the network is likely to balance the nodes' energy consumption, and thus, extend its lifetime. On the other hand, if a given deployment generates topologies that resembles a heavy tail random distribution, few nodes will be more likely to deplete their batteries quickly, and thus, decreases the network lifetime. It is also noticeable that nodes that present higher SBet metric are critical for the network connectivity. It means that if all nodes with higher SBet die, the network will be more likely disconnected.

2.3 Algorithms for wireless sensor networks

Besides the usage of the SBet metric in the design space of WSNs as discussed in Section 2.2, the SBet metric is also useful in the operation space. In order to use this information, the SBet metric of each node must be calculate in a distributed fashion. Ramos et al. (2014) presents a distributed algorithm that is able to calculate the SBet value of each node by using only $O(n)$ messages. This distributed algorithm may be used by different algorithms and protocols that can take advantage of the SBet value.

The main goal of data-collection algorithms is to provide a routing infrastructure to deliver the sensed data through paths that minimize the number of retransmissions and potentially decrease the energy consumption. Moreover, the network may use the processing capacity of its nodes to perform some additional in-network operations. A common task is data fusion (Nakamura et al., 2007) that aims at taking advantage of the data redundancy, increasing data accuracy, reducing the communication load, and saving energy. Krishnamachari et al. (2002) show that a Steiner tree comprised of nodes that detected the events, the sink, and additional nodes creates the optimal infrastructure data-collection infrastructure (Steiner nodes) for this application. The proposed metric, SBet, was also applied to

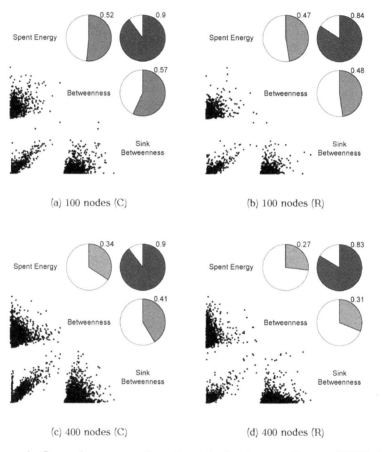

(a) 100 nodes (C)

(b) 100 nodes (R)

(c) 400 nodes (C)

(d) 400 nodes (R)

Figure 4: Corrrelograms and scatterplots for tree routing and URP deployment, 100 and 400 nodes, centered (C) and randomly (R) placed sink

this kind of scenario, in order to build data collection infrastructures that favor the data fusion for event-driven WSNs (Oliveira et al., 2010).

SBet is similar to betweenness in the sense that it represents the centrality in terms of shortest paths, with added nifty properties for the WSN context. For example, SBet is much cheaper to calculate than betweenness. Oliveira et al. (2010) presents a distributed algorithm that uses only $2n$, messages, with n representing the number of nodes, to calculate SBet in non-weighted graphs. Although it is necessary to perform two floods, commonly, shortest path based routing algorithms usually require one flooding to set up the routing structure (Nakamura et al., 2009). Thus, the first flooding can piggyback the necessary data, and only another flooding is necessary to complete the algorithm. To calculate the betweenness by a similar approach, $2n^2$ messages are needed, which might be excessively costly for large WSNs.

Moreover, SBet provides a much better representation of the traffic pattern in WSNs than betweenness. For instance, Figure 5 shows a panorama of the distribution of betweenness and SBet. In this figure, the nodes are randomly distributed on the sensor field. The gray level of the nodes is proportional to its betweenness or SBet. The greater the betweenness or SBet are, the darker the point is. The sink node is represented by a triangle. When the sink is positioned at the center of the network, as shown in Figures 5a and 5b, both metrics are able to distinguish the nodes that concentrate more routes toward the sink. SBet is more selective and presents high values only in nodes that, in fact, participate in more paths to the sink. Betweenness presents more nodes with high values far from the sink, once it considers paths among all nodes. When the sink is located at the corner (Figures 5c and 5d), betweenness fails to represent nodes that participate in more paths to the sink, and lacks the ability of characterizing the traffic pattern of WSNs, while SBet maintains this desirable ability.

The SBet ability of characterizing the traffic pattern of WSNs can be used for many purposes. For instance, on the one hand, the nodes with high SBet are more likely to be data fusion points, since they concentrate more paths toward the sink. On the other hand, those nodes are more likely to deplete their energy quickly due to the same reason, i.e., they concentrate more paths toward the sink. Thus, nodes with high SBet can be preferred or avoided in a data forward path toward the sink, depending on the scenario. In event-driven data fusion scenarios, the higher the SBet of a node, the better it is to participate in the data-collection infrastructure since it is a good candidate to be a data fusion point. For continuous data scenarios, SBet can be used to balance the relay task among sensors that belong to the same neighborhood.

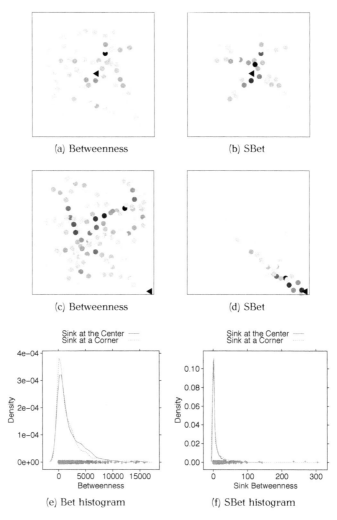

Figure 5: Examples of Betweenness and SBet values for two sink positions, center and corner and their respective histograms (the sink is represented by the triangle)

Another important property of SBet in the context of the energy hole problem is its correlation with the energy spent by the nodes. As the transmission is usually the task that spends more energy, a metric that is able to indicate the nodes that are more likely to transmit more packets is useful to try to balance the workload. We showed in Section 2.2, that the SBet metric is able to characterize the energy consumption in a wide variety of WSNs scenarios. Thus, in the following Sections we show its application in the relay task workload balance, and thus, in the mitigation of the energy hole problem.

The SBet metric was firstly used to balance the relay task workload of a node in a WSN in Ramos et al. (2014).

In this proposal, every node is able to know the SBet value of its neighbor. By using the distributed algorithm to calculate the SBet metric, it is easy to include the calculated SBet value into the packets already used by the protocol. Thus, with no additional cost, all nodes can be aware of their neighbors SBet. Therefore, every time a node has a packet to transmit (it can be its own packet or a relay packet), the node can use the SBet of its neighbors to choose which neighbor will be used to relay the packet. The hypothesis that was verified is that making this choice in a per packet basis and balancing the load among its neighbors guided by the SBet lead to distribute the workload among neighbors as even as possible. As a node with high SBet is more likely to be used as a relay by its neighbors, the routing algorithm uses the SBet to decrease the probability of those nodes to be chosen, and thus, balance the load among the neighbors.

The decision rule used to choose the next hop of node $v_i \in V$ is performed as follows:

1. **SBet calculation and announcement**: we apply the distributed algorithm described in Oliveira et al. (2010) to calculate the SBet, and piggyback its value (announcement) into the packet of the second flooding. The node must calculate its own SBet before sending this packet.

2. **Neighbor filtering**: to ensure the creation of the shortest paths, only neighbors that are closer to the sink than node v_i are eligible as relays, where Ω_i is the set of eligible nodes. The algorithm used in the last phase provides this information.

3. **SBet normalization**: SBet is normalized based on the set Ω_i. This metric is called nSBet (normalized SBet). We observe that each neighborhood presents different values of SBet, and, thus, this process makes the values comparable.

4. **Neighborhood probability assignment:** a probability inversely proportional to nSBet is assigned to every eligible node $v_j \in \Omega_i$. This is the probability of node v_j be chosen as relay of node v_i.

5. **Relay selection:** the relay node is randomly selected according to the probability of item 4.

To define the probabilities of item 4, we use a parameter called "temperature" that controls how intensely the node v_i will try to avoid the choice of a neighbor with high SBet. The probability of node $v_i \in V$ to choose node $v_j \in \Omega_i$ as its relay at temperature T is:

$$\Pr{}_T^i(j) \propto e^{-\mathrm{nSBet}(j)/T}, \tag{4}$$

thus, we calculate

$$k_T^i = \sum_{j \in \Omega_i} e^{-\mathrm{nSBet}(j)/T}, \tag{5}$$

and use (5) to transform the values of (4) in probabilities, as follows:

$$\Pr{}_T^i(j) = \frac{1}{k_T^i} e^{-\mathrm{nSBet}(j)/T}. \tag{6}$$

Equation (6) represents the Boltzmann distribution[1] with the Boltzmann's constant equal to one. Thus, when $T \to \infty$ all relay candidates have the same probability to be chosen, i.e., the process approximates to an uniform random distribution. When $T \to 0$, the node with the smallest SBet value will be chosen with high probability. Both situations are not desirable for our problem.

By using this routing algorithm, it was shown that the relay task workload could be more evenly distributed among the nodes and, thus, prolonging the network lifetime.

The SBet metric can also be used in other scenarios such as in media access control protocols (MAC). They can benefit from the SBet knowledge to plan the WSNs nodes' duty-cycle and potentially increase the energy savings. This application is left as a future work direction.

3 Final remarks

The wireless networks are expected to form the foundation of an intelligent society where people and equipment can be connected anywhere, anytime

---

[1]Also known as the Gibbs Distribution (Kirkpatrick et al., 1983).

with anything. A high degree of connectivity will be a key enabler for new innovative technologies and applications that can benefit from information sharing. In that context, wireless sensor networks will play a key role in obtaining data from different contexts that can eventually be transformed into knowledge.

This vision, however, can only be achieved if high quality research supplies new ideas for improved, affordable solutions and if there are highly qualified researchers to propose these innovations. Frery's work in that direction, starting with the topology design of wireless sensor networks, leading to the study of a more appropriate centrality metric for such networks and the design of a distributed algorithm for such a metric, is very elegant, solid, and has contributed to push our knowledge about the design of wireless sensor networks even further.

References

Akyildiz, I. F., Su, W., Sankarasubramaniam, Y. & Cayirci, E. (2002), 'Wireless Sensor Networks: A Survey', *Computer Networks* **38**(4), 393–422.

Baddeley, A. & Turner, R. (2005), 'Spatstat: An R Package for Analyzing Spatial Point Patterns', *Journal of Statistical Software* **12**(6), 1–42.

Baddeley, A. J. (2007), Spatial Point Processes and their Applications, *in* W. Weil, ed., 'Lecture Notes in Mathematics: Stochastic Geometry', Vol. 1892 of *Lecture Notes in Mathematics: Stochastic Geometry*, Springer Verlag, Berlin.

Beauchamp, M. A. (1965), 'An Improved Index of Centrality', *Behavioral Science* **10**(2), 161–163.

Bonacich, P. (1972), 'Factoring and Weighting Approaches to Status Scores and Clique Identification', *Journal of Mathematical Sociology* **2**(1), 113–120.

Brin, S. & Page, L. (1998), 'The Anatomy of a Large-Scale Hypertextual Web Search Engine', *Computer Networks and ISDN Systems* **30**(1), 1–7.

Costa, L., Rodrigues, F. A., Travieso, G. & Boa, P. R. V. (2007), 'Characterization of Complex Networks: A Survey of Measurements', *Advances in Physics* **56**(1), 167–242.

Cui, J.-H., Kong, J., Gerla, M. & Zhou, S. (2006), 'The Challenges of Building Mobile Underwater Wireless Networks for Aquatic Applications', *IEEE Network* **20**(3), 12–18.

Culler, D., Estrin, D. & Srivastava, M. (2004), 'Guest Editors' Introduction: Overview of Sensor Networks', *Computer* **37**(8), 41–49.

Freeman, L. C. (1977), 'A Set of Measures of Centrality Based on Betweenness', *Sociometry* **40**(1), 35–41.

Freeman, L. C. (1978–1979), 'Centrality in Social Networks Conceptual Clarification', *Social Networks* **1**(3), 215–239.

Frery, A. C., Ramos, H., Alencar-Neto, J. & Nakamura, E. (2008), Error Estimation in Wireless Sensor Networks, *in* 'Proceedings of the 2008 ACM Symposium on Applied Computing (SAC '08)', pp. 1923–1928.

Frery, A. C., Ramos, H. S., Alencar-Neto, J., Nakamura, E. & Loureiro, A. A. F. (2010), 'Data Driven Performance Evaluation of Wireless Sensor Networks', *Sensors* **10**(3), 2150–2168.

Hage, P. & Harary, F. (1995), 'Eccentricity and Centrality in Networks', *Social Networks* **17**(1), 57–63.

Kirkpatrick, S., Gelatt, C. D. & Vecchi, M. P. (1983), 'Optimization by Simulated Annealing', *Science* **220**(4598), 671–680.

Kleinberg, J. M. (1999), 'Authoritative Sources in a Hyperlinked Environment', *Journal of the ACM* **46**(5), 604–632.

Krishnamachari, B., Estrin, D. & Wicker, S. (2002), The Impact of Data Aggregation in Wireless Sensor Networks, *in* 'Proceedings of the 22nd International Conference on Distributed Computing Systems Workshops (ICDCSW '02)', pp. 575–578.

Nakamura, E. F., Loureiro, A. A. F. & Frery, A. C. (2007), 'Information Fusion for Wireless Sensor Networks: Methods, Models, and Classifications', *ACM Computing Surveys* **39**(3), 9/1–9/55.

Nakamura, E. F., Ramos, H. S., Villas, L. A., de Oliveira, H. A. B. F., de Aquino, A. L. L. & Loureiro, A. A. F. (2009), 'A Reactive Role Assignment for Data Routing in Event-Based Wireless Sensor Networks', *Computer Networks* **53**(12), 1980–1996.

Oliveira, E. M. R., Ramos, H. S., & Loureiro, A. A. F. (2010), Centrality-based Routing for Wireless Sensor Networks, *in* 'Proceedings of the 2010 IFIP Wireless Days (WD '10)', pp. 1–5.

Ramos, H. S., Boukerche, A., Frery, A. C. & Loureiro, A. A. F. (2014), 'Topology-Aware Design of Wireless Sensor Networks', ACM Transactions on Sensor Networks. (in submission).

Ramos, H. S., Guidoni, D., Boukerche, A., Nakamura, E. F., Frery, A. C. & Loureiro, A. A. F. (2011), Topology-related Modeling and Characterization of Wireless Sensor Networks, in 'Proceedings of the 8th ACM Symposium on Performance Evaluation of Wireless Ad Hoc, Sensor, and Ubiquitous Networks (PE-WASUN '11)', pp. 33–40.

Ramos, H. S., Oliveira, E. M. R., Boukerche, A., Frery, A. C. & Loureiro, A. A. F. (2012), Characterization and Mitigation of the Energy Hole Problem of Many-to-One Communication in Wireless Sensor Networks, in 'Proceedigns of the 2012 International Conference on Computing, Networking and Communications (ICNC '12)', pp. 954–958.

Sabidussi, G. (1966), 'The centrality of a Graph', *Psychometrika* **31**(4), 581–603.

Wu, X., Chen, G. & Das, S. K. (2008), 'Avoiding Energy Holes in Wireless Sensor Networks with Nonuniform Node Distribution', *IEEE Transactions on Parallel and Distributed Systems* **19**(5), 710–720.

Yarvis, M., Kushalnagar, N., Singh, H., Rangarajan, A., Liu, Y. & Singh, S. (2005), Exploiting Heterogeneity in Sensor Networks, in 'Proceedings of the 24th Annual Joint Conference of the IEEE Computer and Communications Societies (INFOCOM '05)', Vol. 2, pp. 878–890.

Younis, O., Krunz, M. & Ramasubramania, S. (2006), 'Node Clustering in Wireless Sensor Networks: Recent Developments and Deployment Challenges', *IEEE Network* **20**(3), 20–25.

Evaluation of a clustering approach for the colorectal cancer prognosis

Felipe Prata Lima*[†]   Eliana Silva de Almeida[†]

Manoel Alvaro de Lins Freitas Neto[‡]

André Atanasio Maranhão Almeida[§]

Felipe José de Queiroz Sarmento[†]

* Instituto Federal de Alagoas – Campus Arapiraca
felipepratalima@gmail.com

[†] Laboratório de Computação Científica e Análise Numérica - Instituto de
Computação (LaCCAN/IC/UFAL)
{eliana.almeida,felipesarmento}@gmail.com

[‡] Faculdade de Medicina (FAMED/UFAL)
mlinsneto@gmail.com

[§] Instituto Federal da Paraíba – Campus Cajazeiras
andre.atanasio@gmail.com

1   Introduction

Colorectal cancer (CRC) is one of the most incident cancers in the world. It was estimated 136,830 new cases and 50,310 deaths in the United States in 2014 – about 8% and 8.5% of estimated new cases and deaths from the cases considering all cancer types in the country (Siegel et al., 2014).

In Brazil, considering deaths by cancer, about 5.8% of these deaths is due to colorectal cancer (INCA, 2012). In the world, CRC is considered the third type of cancer in males and second in females and the fourth with more death cases in males and the third in females.

Prognostic is a prediction about the development of the disease. In cancer cases, the main prognostic values are rates of relapse after completion of treatment and patient survival after diagnosis through time (Clark et al., 2003; Moons et al., 2009).

Prediction is essential to practice of medicine. In many types of cancer, prognostic models are used to do this prediction. They are known by staging systems and use information about cancer anatomic extension as the main predictor to patient survival. Tumor-Node-Metastasis (TNM) Staging System, supported by American Joint Committee on Cancer (AJCC) and Union for International Cancer Control (UICC), is the main staging system model used for CRC. In this model, the evaluation of CRC is obtained by anatomical and pathological characteristics.

Recent works indicate limitations on TNM Staging System (Greene & Sobin, 2008; MM Center et al., 2009). They consider this model very simple once it has only the staging as predictor. Research has been conducted to find new prognostic models and new methodologies to improve these models. Some of them focus on multiple predictors, by using clinical, demographics, economics, social or biological variables and multivariate approaches to analyze data and construct models (Zhang et al., 1997; Birkenkamp-Demtroder et al., 2002; Chang et al., 2009; Chen et al., 2009; Walther et al., 2009; Wang et al., 2011; Marisa et al., 2013; Steyerberg et al., 2013).

In this study we propose a prognostic model to CRC, based on the work of Xing et al. (2007), that considers a distinguished data clustering approach applied on breast cancer. The approach proposed by Xing et al. (2007) was used to evaluate and validade CRC cancer data. A characterization of a prognostic classification for CRC survival was obtained. In the next section some concepts related to colorectal cancer is presented. Data clustering is defined In Section 3. Following, in Section 4, the method used to develop this work is depicted. Results and conclusions can be seen in sections 5 and 6, respectively.

2   Colorectal cancer

Cancer is a disease characterized by the uncontrolled reproduction of abnormal cells, by a process called oncogenesis. All cancers begin in cells, the body's basic unit of life. To understand cancer, it is helpful to know what happens when normal cells become cancer cells.

The body is made up of many types of cells. These cells grow and divide in a controlled way to produce more cells and to keep the body healthy. When cells become old or damaged, they die and are replaced for new cells. However, sometimes this orderly process goes wrong. The genetic material (DNA) of a cell can become damaged or changed, producing mutations that affect normal cell growth and division. When this happens, cells do not die when they should do and new cells appear, despite the body does not need them. The extra cells may form a mass of tissue called a tumor. Cancer or neoplasia is a term used for diseases in which abnormal cells divide without control and these cells can spread to other parts of the body through the blood and lymph systems.

There are more than 100 different types of neoplasias. Most of them are named for the organ or type of cell in which they start - for example, cancer that begins in the colon and rectum are called colorectal cancer (CRC) (INCA, 2012).

Colorectal cancer (CRC) is one of the most common cancers in developed countries, and its incidence has been continuously increasing. Even though it is a well established model of carcinogenesis, it is still an important cause of mortality, affecting approximately 782 thousand people throughout the world each year (Fearon & Vogelstein, 1990).

Survival rates for CRC are considered good if the disease is early diagnosed. The global mean survival in 5 years is around 55% in developed countries and 40% in developing countries. With this relatively good prognosis, CRC is the second most prevalent type of cancer in the world, with approximately 2.4 million diagnosed living people, coming after breast cancer among women. Estimations of the National Cancer Institute (INCA) in 2010 reported that the number of new CRC cases in Brazil was 32,600, being 15,070 among men and 17,530 among women[1].

The carcinoembryonic antigen (CEA) is the most used method of prognostic evaluation to follow-up patients with CRC. Some studies have evaluated the prognostic value of CEA serum quantification, correlating it to established morphological variables. These are represented by the different staging forms and demonstrate the association between high levels of antigen and unfavorable prognosis. However, results are controversial (Rocha et al., 2012).

The attempts that are currently used for the prognostic evaluation have some limitations, since they are restricted to the observation of tumor morphology, as observed in the TNM staging. This system is the most used tool to classify malignant tumors. The description of its anatomical extension is provided by the evaluation of tumor aggressiveness and invasibility. The tumor, lymph nodes and metastasis classification (TNM) is currently used for the postoperative staging and prognosis. In this model, T (TX, T0, Tis, T1-T4) describes invasion degrees of primary tumor in bowel wall; N (NX, N0-N3) considers the presence or absence of tumor invasion on lymph nodes in the region; and M (MX, M0, M1) indicates the presence or absence of tumor invasion in another organs (metastasis). These values

---

[1]http://www.inca.gov.br/estimativa/2012

are combined to give a staging (I, II, III or IV). The staging of tumor is fundamental to define strategies for patient health care (Horton & Tepper, 2005; Greene & Sobin, 2008).

Patients with colorectal cancer have a more favorable prognosis when the tumor samples revealed less involvement of lymph nodes, not invaded blood or lymph atic vessels neither perineural invasion. Other indicators of poor prognosis for patients undergoing a complete extirpation include a poorly differentiated histology, tumor adherence to adjacent organs, intestinal perforation and colonic obstruction at diagnosis.

3 Data clustering

Data clustering can be defined as a set of procedures that identifies similarity of objects for multivariate data analysis. Homogeneous groups of objects can be revealed or described by similarity forming data sets(Jain et al., 1999).

Briefly, the process for data clustering consists of the following steps:

Feature selection/extraction: in this step, the features to be used in separation of objects in groups are identified. Selection refers to a features choice from a set of possibles features. Extraction refers to transform selected features to better represent the objects.

Construction/definition of a Data Clustering Algorithm: this algorithm uses a proximity measure and a criteria function. The proximity measure describes the similarity or dissimilarity between pairs of objects in the sample. The criteria function resumes the data clustering algorithm to an optimization problem.

Clustering validation: this step consists of a validation test for the selected and extracted features.

Results interpretation: the results are analyzed by the specialists to give them the correct meaning (Xu & Wunsch, 2009).

Data in clustering algorithms are usually presented by two ways: a n-by-p matrix or a n-by-n matrix. The first one presents a set of n objects from a sample, where p variable values were measured. A n-by-n matrix, also known by proximity matrix, presents proximity values between pairs of objects in the sample. Usually they can be defined by dissimilarity measures (distance), meaning how distinguished are the objects. Proximity values between objects in a sample can be presented by similarity measures. These measures suggest how similar are pairs of objects. Usually the Euclidean distance is used as a proximity measure to construct dissimilarity matrix when the object sample contains only continuous variables (Jain et al., 1999):

$$d(x_i, x_j) = [\sum_{k=1}^{p}(x_{ik} - x_{jk})^2]^{1/2}. \tag{1}$$

Data clustering usually is classified as **Partitional** or **Hierarchical**, according to the structure that is produced by their algorithm. In the Partitional cluster, the results produce distinguished groups with no relation between them. There is only one partition of data. The Hierarchical cluster, a set of nested partitions obeying a tree data structure are produced as result. The partitions are defined by the branches of the tree.

3.1 Partitional clustering

In partitional clustering, n objects sample is divided in k groups, such that: (i) each group (cluster) contains at least one object; (ii) each object belongs to a group. A challenge to use this method is to define the initial number of clusters. Finding the best cluster requires several executions of the method for different numbers of cluster and, finally, choose the best one. (Kaufman & Rousseeuw, 1990). Usually, partitional clustering methods adopt a criteria function to optimize. The most common is the mean square error:

$$MSE = \sum_{j=1}^{K}\sum_{i=1}^{n_j}\|\mathbf{x}_i^{(j)} - \mathbf{c}_j\|^2, \tag{2}$$

where $\mathbf{x}_i^{(j)}$ is the ith object indexed by j and $\mathbf{c}_j$ is the group centroid. A known method based on this criteria function is the k-means algorithm (Jain et al., 1999), whose steps are:

1. Select k centroids randomly.

2. Assign each object to the nearest centroid.

3. Recalculate the centroids.

4. Repeat steps 2 and 3 until no reassignment of objects occurs

Centroids represent the means of the groups, obtained from the objects in the sample. Another algorithm, more robust than k-means, is the

Partitioning Around Medoids (PAM). PAM algorithm uses medoids, that is real objects from the sample (Kaufman & Rousseeuw, 1990; Xu et al., 2005).

3.2 Hierarchical clustering

Hierarchical clustering can be classified considering how hierarchy is constructed, as:

Agglomerative: in this case, initially, each object form a group. Then, fusions of groups occur successively until achieving a certain criterion.

Divisive: alternatively, from a group with all objects, divisions occur, until achieving a certain criterion.

The resulting hierarchy may be represented by a dendrogram structure that allows to visualize nesting between the groups and their levels of similarity. So, different groups can be visualized by dividing the dendrogram at different levels.

Hierarchical clutering methods can be described by following steps:

1. Construct N groups, each one, with one object.

2. Use the proximity matrix to find the least distance between two groups and perform a fusion of them.

3. Update the proximity matrix.

4. Repeat steps 2 and 3 until there is only one group.

Linkage methods, that is the way to define proximity between groups, can distinguish hierarchical clustering methods. The most known linkage methods are:

Single linkage: describes the groups proximity from the least distance between two objects of each group.

Complete linkage: alternatively to single linkage, describes the groups proximity from the distance between two most distant objects of each group.

Average linkage: the group proximity is described by the mean of distance between all pairs of objects, where each pair is formed by objects of different groups.

Table 1: Fields and their position in the SEER data files.

Field	Position
Patient ID number	[1..8]
Survival months	[301..304]
Site Recode	[194..198]
SEER modified AJCC Stage 3rd ed (1988-2003)	[239..240]
Grade	[58]
Sex	[24]
Cause of Death to SEER site recode	[255..259]
SEER Cause-Specific Death Classification	[272]

Weighted average linkage: describes the group proximity similarly to average linkage, just considering a weight to perform the proximity based on the number of objects in each group.

Centroid linkage: describes the group proximity as a distance between centroids (mean) of groups.

Median linkage: is similar to Centroid linkage, but considers weight given to centroids.

Ward method: minimizes the increase of the sum of the squared errors as criterion for merge groups (Xu & Wunsch, 2009).

The Algorithm for Clustering of Cancer Data (ACCD), explained in Section 4, use the ideas of agglomerative clustering in its approach to generate groups of patients, which are related to cancer prognosis (Xing et al., 2007).

4   Methods

The colorectal cancer cases assessed in this work were obtained from the Surveillance, Epidemiology, and End Results (SEER) Program from National Cancer Institute (NCI) (http://seer.cancer.gov/data/, 1973-2011) (Surveillance, Epidemiology, and End Results Program of the National Cancer Institute SEER-NCI, 2013). The data files from SEER data, are organized as one case per line, and each field or position (or set of positions) in the line represents a value related to the case. Relevant fields for this study are presented in the Table 1.

SEER data have a common characteristic in cancer data: the censoring. Censoring is an indicator of missing information about the patient survival

Table 2: Typical survival dataset. The presence of censoring is characterized by the combination of the Time and Status field, where 0 indicates the
censoring in the respective time, and 1 the occurrence of the event.

ID	Time (T)	Status (δ)	Gender	Age
1	30	0	F	54
2	14	1	F	34
3	23	1	M	65
4	11	1	F	45
5	12	0	M	44

time. The Table 2 presents a typical suvirval dataset. The censoring is
characterized by the Status field. The value 0 indicates that the event
did not occur at least until the time in Time field and no more information
is known. The value 1 indicates that the event occurs at the time in the
Time field. In this case, this is called right-censoring (Carvalho, 2011;
Kleinbaum & Klein, 2005).

The data analysis in this case is performed using survival analysis techniques, a collection of statistical procedures to analyze data, considering the
time until the occurrence of an event (Kleinbaum & Klein, 2005). The main
definition of this field is the survival function:

$$S(t) = Pr(T > t), \tag{3}$$

the probability of the event non-ocurrence for at least a time t (Clark et
al., 2003).

If the censoring is in the data, the Kaplan-Meier estimator can be used
to estimate the survival function:

$$\hat{S}(t_j) = \hat{S}(t_{j-1}) \times \frac{R(t_j) - \Delta N(t_j)}{R(t_j)}, \tag{4}$$

where $R(t_j)$ is the number of patients in risk and $\Delta N(t_j)$ is the number
of event, both in the time t_j (Carvalho, 2011). The plots of $S(t)$ and the
cumulative hazard ($\Delta(t) = -ln(S(t))$), the inverse of the survival function,
are the main graphical tools for the analysis of survival (Clark et al., 2003).
The Figure 1 depicts examples of these plots.

A non-parametric hypothesis test can be performed if there is difference between two or more survival curves for different groups of patients.
The most used test is the log-rank test. It compares observed and expected

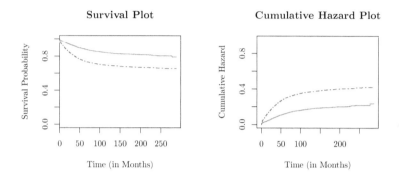

Figure 1:   Survival and cumulative hazard plot examples.

event occurrences for each group:

$$X^2 = \sum_i^k \frac{(O_i - E_i)^2}{E_i}. \tag{5}$$

where, for group i, O_i is the observed number of event, E_i is the number of expected event and k is the number of groups, which is compared to a χ^2 distribution with $k - 1$ degrees of freedom (Clark et al., 2003).

Considering the effect of a set of prognostic factors in the survival analysis, the Cox Proportional Hazards Model is the most used method. It is a regression model, which describes the relation between these factors (covariates) and the incidence of the event as by the risk at a time t. It can be expressed as:

$$\lambda(t) = \lambda_0(t) \, \exp(x_1\beta_1 + x_2\beta_2 + \ldots + x_p\beta_p), \tag{6}$$

where $\lambda_0(t)$ is a baseline hazard, and $(\beta_1, \beta_2, \ldots, \beta_p)$ are coefficients, which measure the impacts of each prognostic factor, for the covariates $(x_1, x_2, \ldots, x_p)$ (Bradburn et al., 2003). As a measure of the goodness-of-fit, the Akaike Information Criterion (AIC) can be computed for this model as:

$$AIC = -2LogL + kp \tag{7}$$

where, $LogL$ is the log-likelihood, k is a constant and p is the number of parameters of the model (Chen et al., 2012; Mallett et al., 2010).

The Algorithm for Clustering of Cancer Data (ACCD) (Xing et al., 2007) was used for grouping CRC survival data. Our aim is proposing a new

Table 3: Possible values to be used in the computation of combinations for these hypothetical variables.

Variables	Values	# of possible values
X_1	{I, II, III, IV}	4
X_2	{I, II, III, IV}	4
X_3	{A, B}	2
X_4	{Y, N}	2

Table 4: Example of combinations based on variables X1, X2, X3 and X4. Each combination is formed by a set of cases for a set of variable values that they can assume.

Combinations	X_1	X_2	X_3	X_4	Objects
x_1	I	I	A	Y	Cases for $\{X_1 = I, X_2 = I, X_3 = A, X_4 = Y\}$
x_2	I	I	A	N	Cases for $\{X_1 = I, X_2 = I, X_3 = A, X_4 = N\}$
x_3	I	I	B	Y	Cases for $\{X_1 = I, X_2 = I, X_3 = B, X_4 = Y\}$
x_4	I	I	B	N	Cases for $\{X_1 = I, X_2 = I, X_3 = B, X_4 = N\}$
x_5	I	II	A	Y	Cases for $\{X_1 = I, X_2 = II, X_3 = A, X_4 = Y\}$
x_6	I	II	A	N	Cases for $\{X_1 = I, X_2 = II, X_3 = A, X_4 = N\}$
...	...	...	...	...	...
x_{64}	IV	IV	B	N	Cases for $\{X_1 = IV, X_2 = IV, X_3 = B, X_4 = N\}$

prognostic classification for this disease based on multiple prognostic factors. This algorithm uses the comcept of Combination. A Combination is a subset of prognostic factors, extracted from a dataset of cancer cases, by combination of some values. In this way, various combinations can be obtained from a dataset. To illustrate these combinations, suppose that X_1, X_2, X_3 and X_4 are prognostic factors and whose set of possible hypothetical values are presented in the Table 3.

It must be observed that all prognostic factors should be categorical. The combinations are obtained according to the possible values for each prognostic factor. In the example 64 combinations were obtained ($4 \times 4 \times 2 \times 2$), as presented in the Table 4.

Each case of the dataset is assigned to one combination. For the ACCD approach, a combination should be interpreted as an object in the sense of data clustering.

The ACCD algorithm used in this work is described as follows:

---

1. Compute the initial set of combinations.
2. Compute the p-value of a statistical test for each pair of combinations.

3. Perform the fusion of each pair of combinations which presents the greater p-value and $p > 0.05$, forming a new combination, and repeat the step (2). When all the pairs of combination present a p-value ≤ 0.05, stop.

We applied the same test used in Xing et al. (2007) to compute the statistic of the step 2, the log-rank test.

5 Results

The prognostic factors used in this study are: TNM Stage, Grade, Site and Gender. All data were extracted from SEER database, considering 138,806 cases of colorectal adenocarcinoma with diagnosis between 1998 and 2003. Initially, For these 4 prognostic factors, a total of 96 initial combinations were obtained ($4 \times 4 \times 3 \times 2$). Then, 23 combinations were dropped by have less than 100 cases, and 73 combinations were obtained. One combination was manually dropped from the dataset. It was the only one with the presence of the grade IV appearance. Finally 72 combinations with 136,961 cases were obtained. Table 5 summarizes the prognostic factors distribution of the final dataset of the experiment.

The Figure 2 presents the survival and cumulative hazard graphics for the prognostic factors of the experiment, except for the TNM stage.

Survival analysis techniques were applied to dataset, considering TNM stage as a prognostic factor. As a result, a model for performing comparisons with TNM stage system can be provided. Figure 3 presents the survival and cumulative hazards for these data. Table 6 presents the cox proportional hazards model, and the Table 7 presents the result of the log-rank test and the AIC value.

From the application of the ACCD approach, the data were clustered into 22 groups. The Figure 4 shows the survival and cumulative hazard graphics based on these groups. It can be observed that the assumption of proportional curves has been satisfied and just a few curves get crossed.

Tables 8 and 9 present the cox proportional hazards model, that was fitted based on the groups obtained by the ACCD approach and its log-rank test and AIC value. The AIC value indicates that this model fits better than the one based on the TNM stage ($1,628,027 < 1,631,214$).

Descriptions of groups compositions are presented in the Tables 10 and 11. These descriptions allow the use of these groups as a basis for the prognostic model derived from the ACCD output. That is, given the

Table 5: Frequencies for the prognostic factors of the final dataset of the experiment, with duplicated and incomplete data dropped, and after compute combinations.

Prognostic Factor		#
SEER modified AJCC Stage 3rd ed (1988-2003)		
	I	28,558
	II	37,430
	III	40,306
	IV	30,667
Grade		
	I	12,481
	II	96,515
	III	27,965
Sex		
	Female	67,744
	Male	69,217
Site Recode ICD-O-3/WHO 2008		
	Distal	39,582
	Proximal	56,041
	Rectum	41,338

Table 6: Cox proportional hazards model fitted from the experiment dataset, based on the TNM stage.

Prognostic Factor	Coefficient	Hazard Ratio	Confidence Interval	p-value
TNM Stage				< 0.0001
I	(0.000)	(1.000)		
II	0.860	2.362	(2.290–2.436)	
III	1.442	4.228	(4.106–4.354)	
IV	2.900	18.173	(17.644–18.717)	

Table 7: Log-rank test and AIC for cox proportional hazards model fitted from the dataset, based on the TNM stage, presented in the Table 6. The log-rank test indicates that there is difference between the survival experiences of these groups ($p = 0$). Note that d.f. means degrees of freedom, which is equal to the number of groups minus one.

Description	Value
Log-rank test	81,123 (3 d.f., $p = 0$)
AIC	1,631,214

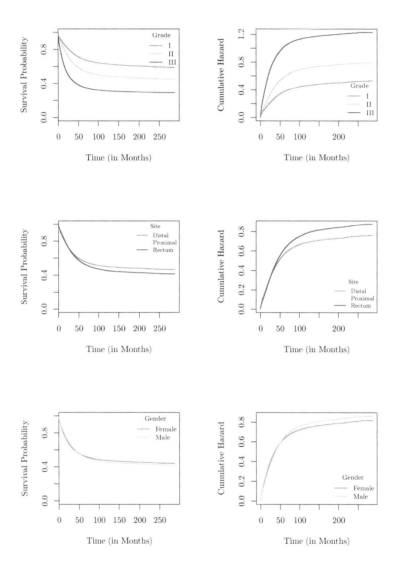

Figure 2: Survival and cumulative hazard graphics from the experiment data. Each graphic presents the survival or cumulative hazard curve by groups based on the different prognostic factors, as described in the legends.

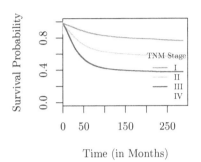 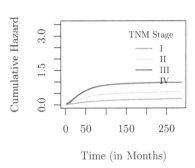

Figure 3:   Survival and cumulative hazard graphics estimated from the dataset. They are presented by groups based on the TNM stage.

Table 8:   Cox proportional hazards model fitted from the experiment dataset, based on the groups obtained by the ACCD approach – 22 groups were presented as the output of the algorithm.

Prognostic Factor	Coefficient	Hazard Ratio	Confidence Interval	p-value
Group				< 0.0001
2	2.434	11.41	(10.45–12.458)	
3	1.125	3.081	(2.779–3.414)	
4	0.728	2.07	(1.894–2.263)	
5	-0.448	0.639	(0.564–0.724)	
6	2.372	10.719	(9.81–11.713)	
7	0.804	2.234	(2.041–2.445)	
8	0.633	1.883	(1.716–2.066)	
9	-0.235	0.791	(0.72–0.868)	
10	2.181	8.857	(8.117–9.665)	
11	0.494	1.639	(1.498–1.792)	
12	1.047	2.849	(2.591–3.132)	
13	-0.607	0.545	(0.491–0.605)	
14	-1.298	0.273	(0.234–0.319)	
15	0.272	1.313	(1.201–1.436)	
16	-1.023	0.36	(0.323–0.4)	
17	-0.809	0.445	(0.373–0.533)	
18	2.116	8.297	(7.564–9.101)	
19	2.731	15.351	(14.036–16.788)	
20	1.239	3.453	(3.155–3.78)	
21	0.877	2.405	(2.202–2.626)	
22	0.092	1.096	(1.002–1.199)	

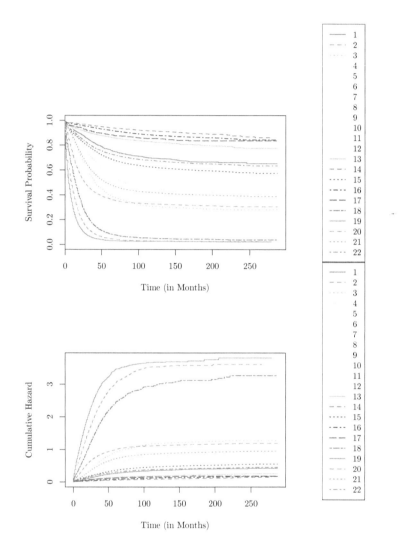

Figure 4:   Survival and cumulative hazard graphics estimated from the experiment data, based on the groups obtained by the ACCD approach – 22 groups were presented as the output result of the algorithm.

Table 9: Log-rank test and AIC for cox propotional hazards model fitted from the dataset, based on the groups obtained by the ACCD approach, presented in the Table 8. The log-rank test indicates that there is difference between the survival experiences of these groups $(p = 0)$.

Description	Value
Log-rank test	86,030 (21 d.f., $p = 0$)
AIC	1,628,027

set of prognostic factors (stage, grade, site and gender) of some patient, it can be found in these tables and the patient's group can be determined. The group has its prognostic characteristics described in the graphics of the Figure 4 and Table 8.

6   Conclusion

This work validates the application of the Algorithm for Clustering of Cancer Data (ACCD) for CRC data, using the SEER data. The results can be characterized as a prognostic classification for the CRC survival. The algorithm demonstrated superior performance when compared to TNM staging system, if we consider the goodness-of-fit using the Akaike Information Criterion (AIC).

Although TNM staging system is supposed to be a robust prognostic factor, the results show that more variables could be incorporated and thereby obtain improvements on survival prediction in case of the colorectal cancer. Furthermore, new methods must be evaluated to deal with the multivariate datasets and provide the knowledge generation about CRC prognosis.

We are evaluating similar methods based on this clustering algorithm, as proposed in Chen et al. (2009). We also will conduct a comparative study of results obtained by the two or more different approaches in order to obtain a better analysis regarding prognosis of colorectal cancer.

7   Acknowledgement

The authors are grateful to Prof. Alejandro C. Frery for his comments and suggestions on statistical analysis. This work was partially sponsored by CNPq, National Council of Technological and Scientific Development - Brazil, and FAPEAL, the Research Foundation of Alagoas - Brazil.

Table 10: Composition for the groups 1 to 11 obtained by the ACCD approach.

Group	Size	Stage	Grade	Site	Gender
1	591	I	III	Rectum	Male
	471	I	III	Rectum	Female
	644	II	I	Proximal	Male
2	1,278	IV	III	Rectum	Male
	874	IV	III	Rectum	Female
	1,178	IV	III	Distal	Male
	4,175	IV	II	Proximal	Female
3	1,606	III	III	Rectum	Male
4	719	II	III	Rectum	Male
	280	III	I	Rectum	Female
	3,769	III	II	Rectum	Female
	4,076	III	II	Distal	Male
5	978	I	I	Rectum	Male
	827	I	I	Rectum	Female
	273	I	III	Proximal	Male
6	262	IV	I	Rectum	Male
	202	IV	I	Rectum	Female
	1,046	IV	III	Distal	Female
	299	IV	I	Proximal	Male
	3,884	IV	II	Proximal	Male
	294	IV	I	Proximal	Female
7	342	III	I	Rectum	Male
	4,741	III	II	Rectum	Male
	577	II	III	Rectum	Female
	426	III	I	Proximal	Male
	482	III	I	Proximal	Female
8	458	II	I	Rectum	Male
	306	II	I	Rectum	Female
	343	III	I	Distal	Female
	4,068	III	II	Distal	Female
9	4,791	I	II	Rectum	Male
	3,966	I	II	Rectum	Female
	384	I	III	Proximal	Female
10	3,308	IV	II	Rectum	Male
	2,173	IV	II	Rectum	Female
	235	IV	I	Distal	Male
	3,364	IV	II	Distal	Male
	195	IV	I	Distal	Female
11	4,360	II	II	Rectum	Male
	3,287	II	II	Rectum	Female
	555	II	III	Distal	Male
	321	III	I	Distal	Male
	635	II	III	Distal	Female
	496	II	I	Distal	Female

Table 11: Composition for the groups 12 to 22 obtained by the ACCD approach.

Group	Size	Stage	Grade	Site	Gender
	1,172	III	III	Rectum	Female
12	1,073	III	III	Distal	Male
	994	III	III	Distal	Female
	319	I	III	Distal	Male
13	2,471	I	II	Proximal	Male
	2,952	I	II	Proximal	Female
14	1,236	I	I	Distal	Male
	1,037	I	I	Distal	Female
	455	II	I	Distal	Male
	4,043	II	II	Distal	Male
15	4,320	II	II	Distal	Female
	1,352	II	III	Proximal	Male
	2,021	II	III	Proximal	Female
	3,401	I	II	Distal	Male
16	2,985	I	II	Distal	Female
	860	I	I	Proximal	Female
17	250	I	III	Distal	Female
	766	I	I	Proximal	Male
18	2,957	IV	II	Distal	Female
19	2,224	IV	III	Proximal	Male
	2,719	IV	III	Proximal	Female
20	2,404	III	III	Proximal	Male
	3,250	III	III	Proximal	Female
21	5,040	III	II	Proximal	Male
	5,919	III	II	Proximal	Female
	5,744	II	II	Proximal	Male
22	737	II	I	Proximal	Female
	6,721	II	II	Proximal	Female

References

Birkenkamp-Demtroder, K., Christensen, L. L., Olesen, S. H., Frederiksen, C. M., Laiho, P., Aaltonen, L. A., Laurberg, S., Sørensen, F. B., Hagemann, R. & Ørntoft, T. F. (2002), 'Gene expression in colorectal cancer', *Cancer Research* **62**(15), 4352–4363.

Bradburn, M. J., Clark, T. G., Love, S. B. & Altman, D. G. (2003), 'Survival analysis part II: multivariate data analysis – an introduction to concepts and methods', *British journal of cancer* **89**(3), 431–436.

Carvalho, M. S. (2011), *Análise de Sobrevivência: teoria e aplicações em saúde*, 2 ed., Fiocruz.

Chang, G. J., Hu, C.-Y., Eng, C., Skibber, J. M. & Rodriguez-Bigas, M. A. (2009), 'Practical application of a calculator for conditional survival in colon cancer', *Journal of Clinical Oncology* **27**(35), 5938–5943.

Chen, D., Xing, K., Henson, D., Sheng, L., Schwartz, A. M. & Cheng, X. (2009), 'Developing prognostic systems of cancer patients by ensemble clustering', *Journal of Biomedicine and Biotechnology* **2009**, 1–7.

Chen, H.-C., Kodell, R., Cheng, K. & Chen, J. (2012), 'Assessment of performance of survival prediction models for cancer prognosis', *BMC Medical Research Methodology* **12**(102), 1–11.

Clark, T. G., Bradburn, M. J., Love, S. B. & Altman, D. G. (2003), 'Survival analysis part I: Basic concepts and first analyses', *British Journal of Cancer* **89**(2), 232–238.

Fearon, E. R. & Vogelstein, B. (1990), 'A genetic model for colorectal tumorigenesis', *Cell* **61**(5), 759–767.

Greene, F. & Sobin, L. (2008), 'The staging of cancer: A retrospective and prospective appraisal', *CA: A Cancer Journal for Clinicians* **58**(3), 180–190.

Horton, J. K. & Tepper, J. E. (2005), 'Staging of colorectal cancer: Past, present, and future', *Clinical Colorectal Cancer* **4**(5), 302 – 312.

INCA (2012), *ABC do câncer : abordagens básicas para o controle do câncer*, 2 ed., INCA.

Jain, A. K., Murty, M. N. & Flynn, P. J. (1999), 'Data clustering: A review', *ACM Computing Surveys (CSUR)* **31**(3), 264–323.

Kaufman, L. & Rousseeuw, P. (1990), *Finding Groups in Data: An Introduction to Cluster Analysis*, John Wiley.

Kleinbaum, D. G. & Klein, M. (2005), *Survival Analysis: A Self-Learning Text*, Springer Science and Business Media, LLC.

Mallett, S., Royston, P., Waters, R., Dutton, S. & Altman, D. (2010), 'Reporting performance of prognostic models in cancer: a review', *BMC Medicine* **8**(21), 1–11.

Marisa, L., de Reyniès, A., Duval, A., Selves, J., Gaub, M. P., Vescovo, L., Etienne-Grimaldi, M.-C., Schiappa, R., Guenot, D., Ayadi, M., Kirzin, S., Chazal, M., Fléjou, J.-F., Benchimol, D., Berger, A., Lagarde, A., Pencreach, E., Piard, F., Elias, D., Parc, Y., Olschwang, S., Milano, G., Laurent-Puig, P. & Boige, V. (2013), 'Gene expression classification of colon cancer into molecular subtypes: Characterization, validation, and prognostic value', *PLoS Medicine* **10**(5), e1001453.

MM Center, Jemal, A., Smith, R. & Ward, E. (2009), 'Worldwide variations in colorectal cancer', *CA: A Cancer Journal for Clinicians* **59**(6), 366–378.

Moons, K. G., Royston, P., Vergouwe, Y., Grobbee, D. E. & Altman, D. G. (2009), 'Prognosis and prognostic research: what, why, and how?', *BMJ* **338**(1), 1–8.

Rocha, V. C., Moreira, R. S., Neto, L. & De Freitas, M. Á. (2012), 'Comparative study between the free DNA in peripheral blood and TNM staging in patients with colorectal cancer for prognostic evaluation in the university hospital of the State of Alagoas', *Journal of Coloproctology (Rio de Janeiro)* **32**(2), 127–131.

Siegel, R., Ma, J., Zou, Z. & Jemal, A. (2014), 'Cancer statistics, 2014', *CA: A Cancer Journal for Clinicians* **64**(1), 9–29.

Steyerberg, E. W., Moons, K. G., van der Windt, D. A., Hayden, J. A., Perel, P., Schroter, S., Riley, R. D., Hemingway, H., Altman, D. G., PROGRESS Group et al. (2013), 'Prognosis research strategy (PROGRESS) 3: Prognostic model research', *PLoS Medicine* **10**(2), e1001381.

Surveillance, Epidemiology, and End Results Program of the National Cancer Institute SEER-NCI (2013), 'SEER Data, 1973-2011'. http://seer.cancer.gov/data/.

Walther, A., Johnstone, E., Swanton, C., Midgley, R., Tomlinson, I. & Kerr, D. (2009), 'Genetic prognostic and predictive markers in colorectal cancer', *Nature Reviews Cancer* **9**(7), 489–499.

Wang, S., Wissel, A., Luh, J., Fuller, C., Kalpathy-Cramer, J. & Thomas, CharlesR., J. (2011), 'An interactive tool for individualized estimation of conditional survival in rectal cancer', *Annals of Surgical Oncology* **18**(6), 1547–1552.

Xing, K., Chen, D., Henson, D. & Sheng, L. (2007), A clustering-based approach to predict outcome in cancer patients, *in* 'Sixth International Conference on Machine Learning and Applications – ICMLA 2007', IEEE, pp. 541–546.

Xu, R. & Wunsch, D. (2009), *Clustering*, Wiley-IEEE Press.

Xu, R., Wunsch, D. et al. (2005), 'Survey of clustering algorithms', *Neural Networks, IEEE Transactions on* **16**(3), 645–678.

Zhang, L., Zhou, W., Velculescu, V., Kern, S., Hruban, R., Hamilton, S., Vogelstein, B. & Kinzler, K. (1997), 'Gene expression profiles in normal and cancer cells', *Science* **276**(5316), 1268–1272.

A misspecification test for beta regressions[‡]

Francisco Cribari-Neto[*]   Leonardo B. Lima[†]

[*] Departamento de Estatística
Universidade Federal de Pernambuco, Brazil
cribari@de.ufpe.br

[†] Departamento de Estatística
Universidade Federal de Pernambuco, Brazil
leostat@gmail.com

1   Introduction

Practitioners oftentimes wish to model random variates that are assume values on the standard unit interval $(0, 1)$, such as rates and proportions. Ferrari & Cribari-Neto (2004) proposed a beta regression model which allows such a modeling to be conditioned on a set explanatory variables. To that end, they have used an alternative parameterization in which the beta density is indexed by mean and precision parameters. In their model, the mean of the response is linked to a linear predictor that involves regressors and unknown regression parameters through a link function.

We note that there are alternative specifications for beta regressions; see, e.g., Kieschnick & McCullough (2003), Paolino (2001), Vasconcellos & Cribari-Neto (2005). In what follows we shall work with the beta regression model of Ferrari & Cribari-Neto (2004) for a few reasons. First, in their model the regression structure is placed on the mean response and not on the two parameters that index the beta distribution, which is a more natural approach. Second, their model is similar to the well known class of generalized linear models. Third, the model is defined for several link functions. Fourth, the authors have proposed an algorithm for choosing the initial values of the parameters in the nonlinear optimization scheme used for maximum likelihood estimation and also standard diagnostic measures. Finally, their model has been implemented into a piece of statistical software: the betareg package for the free software R, which is available for download at http://www.R-project.org (Cribari-Neto & Zeileis, 2010); it is thus widely available to practitioners.

Our chief goal in this paper is to present and numerically evaluate the finite-sample performance of a misspecification test for beta regressions. The null hypoth-

[‡]The authors gratefully acknowledge partial financial suport from CNPq and CAPES.

esis under test is that the model is correctly specified and the alternative hypothesis is that it is misspecified. Misspecification can be due to neglected nonlinearities, omitted variables, incorrectly chosen link functions, etc. The proposed test is thus quite useful in empirical applications, since it is able to detect several forms of model misspecifications. When the null hypothesis is rejected, the practitioner must then turn to more specific tests in order to locate the source of the problem. Our test draws on Ramsey's RESET test (Ramsey, 1969), which was devised for the linear regression model. Unlike the standard RESET test, the beta regression RESET testing inference is based on asymptotic (i.e., approximate) critical values. The numerical evidence we report, however, shows that the approximation can be quite accurate when the sample size is not too small.

As a motivating example, consider the dataset collected by Prater (1956). The dependent variable is the proportion of crude oil converted to gasoline after distilation and fractionation, and the potential covariates are: the crude oil gravity (degrees API), the vapor pressure of the crude oil (lbf/in^2), the crude oil 10% point ASTM (i.e., the temperature at which 10% of the crude oil has become vapor), and the temperature (degrees F) at which all the gasoline is vaporized. There are 32 observations on the response and on the independent variables. It has been noted (Daniel & Wood, 1971) that there are only ten sets of values of the first three explanatory variables which correspond to ten different crudes and were subjected to experimentally controlled distillation conditions. The interest lies in modeling the proportion of crude oil converted to gasoline after distillation and fractionation. Ferrari & Cribari-Neto (2004) have fitted beta regressions to these data using a logit link function. Is the model employed by the authors correctly specified? If not, is it possible to shed some light on the nature of the misspecification? We shall return to this application later.

As a second motivating example, we consider the data analyzed by Smithson & Verkuilen (2006) and Espinheira et al. (2008). The dependent variable (y) are scores on a test of reading accuracy of 44 children, and the regressors are dyslexia versus non-dyslexia status (x_2), nonverbal IQ converted to z-scores (x_3) and an interaction variable (x_4). Participants (19 dyslexics and 25 controls) were recruited from primary schools in Australia. The children ages range from eight years five months to twelve years three months. The independent variable x_2 is a dummy variable which equals 1 when the child is dyslexic and -1 otherwise. The observed scores were linearly transformed from their original scale to the open unit interval $(0, 1)$. Smithson & Verkuilen (2006) and Espinheira et al. (2008) consider a variable dispersion beta regression with logit link. Is their regression model correctly specified? We shall address this issue.

It is important to notice that our interest lies in detecting incorrect model specification, that is, we wish to test whether the model is correctly specified against the general alternative hypothesis that there is *some (unspecified) form* model misspecification. That is, the practitioner need not specify the particular type of model misspecification in the alternative hypothesis. If the test we propose indicates that the model is incorrectly specified, then the practitioner can resort to a series of specific likelihood ratio tests and also to tests of nonnested hypotheses to identify which aspect of his/her model is in specification error. The purpose of our test is not to determine which aspect of the regression model is incorrectly specified, but instead to identify whether there is any model misspecification at all.

The paper unfolds as follows. Section 2 introduces the class of beta regressions. The proposed test is described in Section 3. Section 4 contains numerical evidence on the finite-sample performance of different implementations of the proposed test. The two applications introduced above and a third application that uses a large sample size are presented and discussed in Section 5. Finally, some concluding remarks and directions for future research are collected in Section 6.

2   Beta regression

The random variable Y is said to be beta distributed, denoted $Y \sim \mathcal{B}(p, q)$, if its density function is

$$f(y; p, q) = \frac{\Gamma(p + q)}{\Gamma(p)\Gamma(q)} y^{p-1}(1 - y)^{q-1}, \quad 0 < y < 1, p, q > 0,$$

where $\Gamma(\cdot)$ is the gamma function. The corresponding distribution function is

$$F(y; p, q) = \frac{\Gamma(p + q)}{\Gamma(p)\Gamma(q)} \int_0^y t^{p-1}(1 - t)^{q-1} dt. \tag{2.1}$$

The complete beta function is defined as

$$B(p, q) = \frac{\Gamma(p)\Gamma(q)}{\Gamma(p + q)},$$

and the integral

$$B_y(p, q) = \int_0^y t^{p-1}(1 - t)^{q-1} dt$$

is the incomplete beta function. Note that the distribution function in (2.1) can be written as

$$F(y; p, q) = \frac{B_y(p, q)}{B(p, q)}.$$

Note also that $F(y; p, q) = F(1 - y; q, p)$ since if $Y \sim \mathcal{B}(p, q)$, then $1 - Y \sim \mathcal{B}(q, p)$.

The $\mathcal{B}(p, q)$ rth noncentral moment is

$$
\begin{aligned}
\mu_r = E(Y^r) &= \int_0^1 y^r \frac{\Gamma(p+q)}{\Gamma(p)\Gamma(q)} y^{p-1}(1-y)^{q-1} \\
&= \frac{B(p+r, q)}{B(p, q)} = \frac{\Gamma(p+r)\Gamma(p+q)}{\Gamma(p)\Gamma(p+q+r)} \\
&= \frac{p_{(r)}}{(p+q)_{(r)}},
\end{aligned}
$$

where $k_{(r)} = k \times (k+1) \times (k+2) \times \cdots \times (k+r-1)$. Hence, the mean of $Y \sim \mathcal{B}(p, q)$ is given by

$$
\mathbb{E}(Y) = \frac{p}{(p+q)}; \tag{2.2}
$$

the variance is

$$
\mathrm{var}(Y) = \frac{pq}{(p+q)^2(p+q+1)}. \tag{2.3}
$$

Ferrari & Cribari-Neto (2004) have considered an alternative parameterization of the beta distribution. Let

$$
\mu = p/(p+q) \quad \text{and} \quad \phi = p+q,
$$

i.e.,

$$
p = \mu\phi \quad \text{and} \quad q = (1-\mu)\phi.
$$

Hence, from (2.2) and (2.3),

$$
\mathbb{E}(Y) = \mu \quad \text{and} \quad \mathrm{var}(Y) = \frac{V(\mu)}{1+\phi},
$$

where $V(\mu) = \mu(1-\mu)$. Here, μ is the mean of the random variate and ϕ can be viewed as a precision parameter in the sense that, for fixed μ, the larger ϕ the smaller the variance of Y. The density of Y can now be written as

$$
f(y; \mu, \phi) = \frac{\Gamma(\phi)}{\Gamma(\mu\phi)\Gamma((1-\mu)\phi)} y^{\mu\phi-1}(1-y)^{(1-\mu)\phi-1}, \quad 0 < y < 1,
$$

$0 < \mu < 1$ and $\phi > 0$.

The more general case in which the random variate assumes values in (a, b), where a and b are known constants such that $a < b$, can be easily handled by modeling $(Y-a)/(b-a)$, which assumes values in the standard unit interval.

Let $y_1, \ldots, y_n$ be a random sample such that $y_t \sim \mathcal{B}(\mu_t, \phi)$, $t = 1, \ldots, n$. The beta regression model is defined as

$$
g(\mu_t) = \sum_{i=1}^{k} x_{ti}\beta_i = \eta_t,
$$

where $\beta = (\beta_1, \ldots, \beta_k)^\top$ is a k-vector of unknown regression parameters ($\beta \in \mathbb{R}^k$), η_t is a linear predictor and $x_{t1}, \ldots, x_{tk}$ are (fixed) covariate values ($k < n$). The link function $g : (0; 1) \to \mathbb{R}$ must be strictly monotone and twice differentiable. Some commonly used link functions are:

1. logit:
$$g(\mu) = \log(\mu/(1 - \mu));$$

2. probit:
$$g(\mu) = \Phi^{-1}(\mu);$$

3. complementary log-log:
$$g(\mu) = \log\{-\log(1 - \mu)\};$$

4. log-log:
$$g(\mu) = -\log\{-\log(\mu)\};$$

5. Cauchy:
$$g(\mu) = \tan\{\pi(\mu - 0.5)\}.$$

Figure 1 plots these five link functions. We note that the logit, probit and Cauchy functions are similar when μ is close to 0.5. As μ approaches 1 the complementary log-log function diverges more slowly than the other functions. It is also noteworthy that the Cauchy link function behaves differently from the remaining functions when μ is close to the limits of the standard unit interval. When μ is small, the logit and complementary log-log functions behave similarly; when μ is large, the logit and log-log functions display similar behavior.

The log-likelihood function can be expressed as

$$\ell(\beta, \phi) = \sum_{t=1}^{n} \ell_t(\mu_t, \phi),$$

where

$$\begin{aligned} \ell_t(\mu_t, \phi) &= \log \Gamma(\phi) - \log \Gamma(\mu_t \phi) - \log \Gamma((1 - \mu_t)\phi) \\ &+ (\mu_t \phi - 1) \log y_t + \{(1 - \mu_t)\phi - 1\} \log(1 - y_t). \end{aligned}$$

Note that $\mu_t = g^{-1}(\eta_t)$ is a function of β. Parameter estimation is carried out by numerically maximizing the log-likelihood function. The maximum likelihood estimators are biased in small samples. Analytical and numerical bias correcting schemes were proposed by Ospina et al. (2006).[1]

---

[1] For bias correction strategies under the law in the nonregression case, see Cribari-Neto & Vasconcellos (2002).

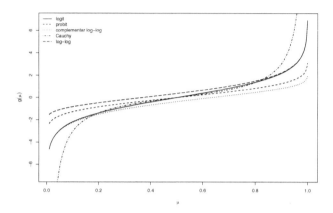

Figure 1: Some link functions.

Fisher's information matrix for (β, ϕ) can be written as

$$K = K(\beta, \phi) = \begin{pmatrix} K_{\beta\beta} & K_{\beta\phi} \\ K_{\phi\beta} & K_{\phi\phi} \end{pmatrix},$$

where $K_{\beta\beta} = \phi X^\top W X$, $K_{\beta\phi} = K_{\phi\beta}^\top = X^\top T c$ and $K_{\phi\phi} = \text{tr}(D)$. Here, X is the $n \times k$ matrix of covariate values, $T = \text{diag}\{1/g'(\mu_1), \ldots, 1/g'(\mu_n)\}$, and $W = \text{diag}\{w_1, \ldots, w_n\}$, with

$$w_t = \phi\{\psi'(\mu_t\phi) + \psi'((1 - \mu_t)\phi)\} \frac{1}{g'(\mu_t)^2},$$

$\psi'(\cdot)$ being the trigamma function, i.e., the first derivative of the digamma function. Also, $D = \text{diag}(d_1, \ldots, d_n)$, with $d_t = \psi'(\mu_t\phi)\mu_t^2 + \psi'((1 - \mu_t)\phi)(1 - \mu_t^2) - \psi'(\phi)$, and $c = (c_1, \ldots, c_n)^\top$, with $c_t = \phi\{\psi'(\mu_t\phi)\mu_t - \psi'((1 - \mu_t)\phi)(1 - \mu_t)\}$. Note that, unlike the class of generalized linear models (McCullagh & Nelder, 1989), β and ϕ are not orthogonal. Finally, we note that the model can be easily extended to allow the precision parameter to vary across observations; see Simas et al. (2010).

3   A misspecification test for beta regressions

Our chief goal is to offer a general misspecification test that can be used by practitioners when fitting beta regressions to data. The proposed test is an extension of the misspecification test proposed by Ramsey (1969) for linear regressions. The underlying main idea is that when the model is correctly specified powers of the fitted values or powers of the regressors should not provide additional contribution to the quality of the fit.

Consider the beta regression model written in matrix form as

$$g(\mu) = X\beta,$$

where μ is the n-vector of mean responses. The proposal is to test the null hypothesis $\mathcal{H}_0 : \theta = 0$ (the model is correctly specified) against the alternative hypothesis $\mathcal{H}_1 : \theta \neq 0$ (the model is not correctly specified), where 0 denotes an $s \times 1$ vector of zeros, when the model is augmented as

$$g(\mu) = X\beta + Z\theta,$$

Z being an $n \times s$ matrix of testing variables and θ is an s-vector of parameters. Rejection of the hypothesis that θ equals zero is taken as evidence that the beta regression at hand is not correctly specified. Note that when θ is not zero, the testing variables used to augment the model do contribute to improving the regression fit. Thus, there is neglected useful information.

The testing variables can be taken to be: (i) powers of the predicted values $\hat{\mu}_t$, (ii) powers of the fitted linear predictor $\hat{\eta}_t$, or (iii) powers of the regressors x_{ti}, $i = 1, \ldots, k$.

In the linear regression model tests on several regression coefficients can be carried out in exact fashion using the F test. Nonetheless, in beta regressions it is necessary to use a test based on a first order asymptotic approximation. Three tests that can be used to test the exclusion of the variables used to augment the model are the likelihood ratio, score and Wald tests. Note that the score test only requires the estimation of the model under the null hypothesis, whereas the Wald test requires the estimation of the unrestricted model and the likelihood ratio test needs information from the two fits, i.e., from the restricted and unrestricted model fits.

Let v denote the complete $(k + s + 1) \times 1$ parameter vector, i.e., $v = (\theta^\top, \beta^\top, \phi)^\top$. We wish to test $\mathcal{H}_0 : \theta_1 = \theta_2 = \cdots = \theta_s = 0$ versus a two-sided alternative hypothesis. Let $\hat{v} = (\hat{\theta}^\top, \hat{\beta}^\top, \hat{\phi})^\top$ and $\tilde{v} = (0^\top, \tilde{\beta}^\top, \tilde{\phi})^\top$ denote the unrestricted and restricted maximum likelihood estimators of v, respectively.

The log-likelihood RESET test statistic is

$$\xi_1 = 2\{\ell(\hat{v}) - \ell(\tilde{v})\},$$

where $\ell(v)$ is the log-likelihood function. The score RESET test statistic can be written as

$$\xi_2 = \tilde{U}_{1\theta}^\top \tilde{K}_{11}^{\theta\theta} \tilde{U}_{1\theta},$$

where $U_{1\theta}$ is the s-vector with the first s elements of the score function for the regression parameters, $K_{11}^{\theta\theta}$ is the $s \times s$ matrix containing the first s rows and the first

s columns of K^{-1}, and tildes denote evaluation at the restricted maximum likelihood estimator. Finally, the Wald RESET test statistic is given by

$$\xi_3 = \hat{\theta}^\top \left(\hat{K}_{11}^{\theta\theta} \right)^{-1} \hat{\theta},$$

where $\hat{K}_{11}^{\theta\theta}$ is $K_{11}^{\theta\theta}$ evaluated at the unrestricted maximum likelihood estimator.

Under $\mathcal{H}_0$, $\xi_j \leadsto \chi_s^2$, $j = 1, 2, 3$, where $\leadsto$ denotes convergence in distribution. Hence, the test can be performed using approximate critical values from the χ_s^2 distribution. The null hypothesis of no model misspecification is rejected at the nominal level $\alpha \in (0, 1)$ when $\xi_j > \chi_{s,\alpha}^2$, $j = 1, 2, 3$, where $\chi_{s,\alpha}^2$ is the upper $1 - \alpha$ quantile of the χ_s^2 distribution.

4 Numerical evaluation

The Monte Carlo simulations shall focus on two important sources of model misspecification, namely: neglected nonlinearities and incorrectly specified link functions. The beta regression model used in the numerical evaluations is

$$g(\mu_t) = \beta_0 + \beta_1 x_{t1} + \beta_2 x_{t2}, \quad t = 1, \ldots, n, \tag{4.1}$$

where $g(\cdot)$ is the link function. The values of the two covariates for a given sample size are kept constant throughout the simulations. The covariate values were selected as random draws from the standard uniform distribution $\mathcal{U}_{(0,1)}$ and the sample sizes considered were $n = 20, 40, 60, 100, 200$. All results are based on 10,000 Monte Carlo replications and all simulations were performed using the Ox matrix programming language (Doornik, 2001). All log-likelihood functions were maximized using the BFGS quasi-Newton nonlinear optimization algorithm (with analytical first derivatives); see Press et al. (1992, Ch. 10) for details on the BFGS method.

In the size simulations, the response was generated according to (4.1). The following link functions were used: logit, probit, complementary log-log and Cauchy. When the link was logit, the parameter values were $\beta_0 = 1, \beta_1 = -1, \beta_2 = 0.8$ and $\phi = 120$. For the remaining link functions, we used $\beta_0 = -1.8, \beta_1 = 1.4, \beta_2 = 1.6$ and $\phi = 120$.

The nonnull simulations were performed in two different scenarios. First, the responses were generated using the following nonlinear specification:

$$\log \left(\frac{\mu_t}{1 - \mu_t} \right) = (\beta_0 + \beta_1 x_{t1} + \beta_2 x_{t2})^\delta, \quad t = 1, \ldots, n,$$

with $\delta = 1.2, 1.4, 1.6, 1.8$. Here, $\beta_0 = 1, \beta_1 = -1, \beta_2 = 0.8$ and $\phi = 120$. Note that the larger the value of δ, the farther away we are from the null hypothesis, since correct model specification would require $\delta = 1$.

The second set of power simulations was intended to evaluate the test's ability to detect incorrectly specified link functions. Estimations were also carried out with the logit link, whereas data generation was performed using the following link functions: (i) probit, (ii) complementary log-log, and (iii) Cauchy. Note that here the link function used in the estimated model is not the correct link function. The true parameter values were set at $\beta_0 = -1.8, \beta_1 = 1.4, \beta_2 = 1.6$ and $\phi = 120$.

We note that all power simulations were performed using exact critical values (estimated from the corresponding size simulations), and not asymptotic (χ^2) critical values. This was done so that powers comparisons would be meaningful in the sense that all tests share the same size.

We have considered the sets of testing variables listed in Table 1. In particular, note that we have considered both powers of $\hat{\mu}_t$ and powers of $\hat{\eta}_t$ since in beta regression, unlike the linear regression model, these quantities are not equal; recall that $\hat{\mu}_t = g^{-1}(\hat{\eta}_t)$.

Table 1: Testing variables.

Powers of fitted values	$\hat{\mu}_t^2$
	$\hat{\mu}_t^3$
	$\hat{\mu}_t^4$
	$(\hat{\mu}_t^2, \hat{\mu}_t^3)$
	$(\hat{\mu}_t^2, \hat{\mu}_t^3, \hat{\mu}_t^4)$
Powers of regressors	(x_{t1}^2, x_{t2}^2)
	(x_{t1}^3, x_{t2}^3)
	(x_{t1}^4, x_{t2}^4)
	$(x_{t1}^2, x_{t1}^3, x_{t2}^2, x_{t2}^3)$
	$(x_{t1}^2, x_{t1}^3, x_{t1}^4, x_{t2}^2, x_{t2}^3, x_{t2}^4)$
Powers of the fitted linear predictor	$\hat{\eta}_t^2$
	$\hat{\eta}_t^3$
	$\hat{\eta}_t^4$
	$(\hat{\eta}_t^2, \hat{\eta}_t^3)$
	$(\hat{\eta}_t^2, \hat{\eta}_t^3, \hat{\eta}_t^4)$

4.1   Size simulations

For brevity, we shall only report simulation results for the logit link based model. The numerical results for the other link functions are similar. Tables 2 to 4 contain the null rejection rates (%) of the beta regression specification test proposed in Section 3 implemented using the score, likelihood ratio and Wald tests, respectively. All tests were carried out at the 10% and 5% nominal levels. The numerical evidence shows that the score implementation of the RESET test consistently outperforms the likelihood ratio and Wald variants. For instance, when $n = 40$ and the tests are based on $\hat{\eta}_t^2$ as the single testing variable, the empirical sizes of the score, likelihood ratio and Wald tests at the 5% nominal level are 5.84%, 6.50% and 7.17%, respectively. When the sample contains only twenty observations, the likelihood ratio and Wald tests can be quite liberal, their empirical sizes (for $\alpha = 5\%$) reaching nearly 25% and 40%, respectively; the size of the most liberal score test in the same setting is 7.24%. Additionally, the finite-sample performances of the likelihood ratio and Wald implementations of the beta regression misspecification test deteriorate when the number of testing variables increases. On the other hand, the score RESET test displays good finite-sample performance for all choices of testing variables, especially when the sample contains at least fourty observations.

4.2   Power simulations: nonlinearity

We shall now move to the first set of power simulations, i.e., simulations where the data generation was done in disagreement with the estimated model. At the outset, our goal is to evaluate whether the proposed test is able to detect misspecification when its source is neglected nonlinearity. For brevity, we only report results on the score implementation of the RESET test, since it clearly outperformed the other two implementations (likelihood ratio and Wald) in the simulations performed under the null hypothesis of correct model specification. The reported results are for the logit link, $\delta = 1.2, 1.4, 1.6, 1.8$ and $\alpha = 10\%, 5\%$. The results for $n = 20, 40, 60, 100$ are displayed in Tables 5 to **??**, respectively. All entries are percentages.

The figures in Tables 5 to 8 show that the choice of testing variables can have a decisive impact on the power of the misspecification test. For instance, when $n = 40$, $\alpha = 5\%$ and $\delta = 1.6$, the estimated powers range from approximately 10% to nearly 70%. It is also noteworthy that the power of the test that uses $\hat{\mu}_t^2$ as testing variable is substantially smaller than that of the test based on $\hat{\eta}_t^2$. It is equally important to note that the use of powers of regressors as testing variables typically leads to nonnegligible power losses. For instance, when $n = 60$, $\alpha = 10\%$ and $\delta = 1.4$, the powers of the tests based on (x_{t1}^2, x_{t2}^2) and $(\hat{\eta}_t^2, \hat{\eta}_t^3)$ are 27.43% and 56.92%, respectively. Finally, we note that the test based on the single testing variable $\hat{\eta}_t^2$ usually outperforms the competition when it comes to detecting neglected nonlinearities.

4.3   Power simulations: link function

We now move to the case where the source of the model misspecification lies in the link function. The model is estimated using the logit link and yet the true data generating process involves a different link function. The results are presented in Tables 9 (probit), 10 (complementary log-log) and 11 (Cauchy). All entries are percentages.

Table 2: Null rejection rates of the RESET score test.

| added regressor(s) | n = 20 | | n = 40 | | n = 60 | | n = 100 | | n = 200 | |
| | α | | α | | α | | α | | α | |
	5%	10%	5%	10%	5%	10%	5%	10%	5%	10%
$\hat{\mu}_t^2$	6.86	13.24	6.02	12.23	5.21	10.46	5.03	10.10	5.31	10.48
$\hat{\mu}_t^3$	6.72	13.06	6.00	11.51	5.27	10.57	5.41	10.77	5.17	10.42
$\hat{\mu}_t^4$	6.80	13.45	5.97	11.70	5.31	10.53	5.28	10.77	5.40	10.37
(x_{t1}^2, x_{t2}^2)	7.18	14.07	5.83	11.48	5.54	11.03	5.14	10.50	5.00	9.93
(x_{t1}^3, x_{t2}^3)	7.10	13.92	5.66	11.59	5.63	10.94	5.11	10.42	4.79	10.04
(x_{t1}^4, x_{t2}^4)	6.94	14.20	5.59	11.63	5.53	11.02	5.14	10.34	4.79	10.12
$\hat{\eta}_t^2$	6.90	13.71	5.84	11.93	5.21	10.72	5.12	10.80	5.51	10.23
$\hat{\eta}_t^3$	7.24	13.40	5.91	11.78	5.23	10.69	5.04	10.62	5.51	10.55
$\hat{\eta}_t^4$	7.00	13.69	5.94	11.95	5.26	10.52	5.23	10.44	5.45	10.64
$(\hat{\mu}_t^2, \hat{\mu}_t^3)$	7.08	13.73	5.88	12.25	5.26	10.88	5.27	10.63	5.31	10.86
$(x_{t1}^2, x_{t1}^3, x_{t2}^2, x_{t2}^3)$	6.40	14.37	5.61	11.61	5.34	10.63	5.20	10.35	5.10	10.42
$(\hat{\eta}_t^2, \hat{\eta}_t^3)$	6.96	13.75	5.99	12.33	5.22	10.79	5.17	10.64	5.40	10.86
$(\hat{\mu}_t^2, \hat{\mu}_t^3, \hat{\mu}_t^4)$	6.61	14.47	5.97	12.11	5.23	10.99	5.34	10.61	4.57	9.53
$(x_{t1}^2, x_{t1}^3, x_{t1}^4, x_{t2}^2, x_{t2}^3, x_{t2}^4)$	5.10	13.76	5.20	11.63	4.92	10.58	5.20	10.60	4.81	10.18
$(\hat{\eta}_t^2, \hat{\eta}_t^3, \hat{\eta}_t^4)$	6.37	13.80	5.65	11.99	5.84	11.58	5.21	10.59	4.83	9.82

Table 3: Null rejection rates of the RESET likelihood ratio test.

added regressor(s)	n = 20 α		n = 40 α		n = 60 α		n = 100 α		n = 200 α	
	5%	10%	5%	10%	5%	10%	5%	10%	5%	10%
$\hat{\mu}_i^2$	8.34	14.76	6.75	12.84	5.60	10.82	5.27	10.41	5.37	10.50
$\hat{\mu}_i^3$	8.05	14.51	6.55	12.27	5.68	11.07	5.77	11.04	5.33	10.51
$\hat{\mu}_i^4$	8.50	15.03	6.69	12.43	5.80	10.87	5.55	11.06	5.46	10.49
(x_{i1}^2, x_{i2}^2)	10.52	17.82	7.14	13.01	6.41	12.05	5.66	11.07	5.15	10.24
(x_{i1}^3, x_{i2}^3)	10.62	17.68	7.01	13.05	6.48	11.88	5.53	11.10	5.06	10.41
(x_{i1}^4, x_{i2}^4)	10.62	17.81	6.94	13.19	6.35	11.86	5.62	10.93	5.01	10.40
$\hat{\eta}_i^2$	8.87	15.24	6.50	12.44	5.72	11.10	5.29	11.00	5.64	10.42
$\hat{\eta}_i^3$	8.91	14.68	6.61	12.40	5.61	11.12	5.24	10.83	5.65	10.61
$\hat{\eta}_i^4$	8.72	14.93	6.77	12.64	5.64	11.16	5.52	10.66	5.59	10.73
$(\hat{\mu}_i^2, \hat{\mu}_i^3)$	10.56	17.45	7.31	13.84	6.23	11.93	5.78	11.28	5.51	11.04
$(x_{i1}^2, x_{i1}^3, x_{i2}^2, x_{i2}^3)$	15.92	25.15	9.00	15.46	7.13	12.98	6.21	11.67	5.56	11.03
$(\hat{\eta}_i^2, \hat{\eta}_i^3)$	10.60	17.41	7.44	13.93	6.21	11.93	5.71	11.15	5.61	11.13
$(\hat{\mu}_i^2, \hat{\mu}_i^3, \hat{\mu}_i^4)$	13.25	21.20	8.40	14.91	6.65	12.80	6.23	11.37	5.00	9.99
$(x_{i1}^2, x_{i1}^3, x_{i1}^4, x_{i2}^2, x_{i2}^3, x_{i2}^4)$	23.70	34.43	10.92	18.10	8.07	14.50	6.82	12.13	5.64	11.28
$(\hat{\eta}_i^2, \hat{\eta}_i^3, \hat{\eta}_i^4)$	12.65	20.60	8.19	14.72	7.29	13.28	6.03	11.50	5.10	10.23

Table 4: Null rejection rates of the RESET Wald test.

added regressor(s)	n = 20		n = 40		n = 60		n = 100		n = 200	
	α		α		α		α		α	
	5%	10%	5%	10%	5%	10%	5%	10%	5%	10%
$\hat{\mu}_i^2$	9.70	16.00	7.39	13.51	6.04	11.28	5.44	10.59	5.53	10.62
$\hat{\mu}_i^3$	9.53	16.05	7.14	13.06	6.02	11.42	6.04	11.21	5.42	10.57
$\hat{\mu}_i^4$	9.91	16.25	7.20	12.88	6.17	11.24	5.86	11.33	5.58	10.66
(x_{i1}^2, x_{i2}^2)	14.02	21.21	8.51	14.44	7.26	12.89	6.17	11.61	5.42	10.51
(x_{i1}^3, x_{i2}^3)	13.92	20.82	8.48	14.39	7.31	12.90	6.05	11.74	5.28	10.63
(x_{i1}^4, x_{i2}^4)	14.00	21.03	8.59	14.44	7.19	12.70	6.22	11.73	5.21	10.68
$\hat{\eta}_i^2$	10.09	16.49	7.17	13.08	6.13	11.56	5.48	11.19	5.72	10.63
$\hat{\eta}_i^3$	10.14	15.88	7.23	13.03	5.97	11.46	5.58	11.10	5.73	10.65
$\hat{\eta}_i^4$	10.06	16.01	7.32	13.14	5.95	11.46	5.54	10.89	5.65	10.78
$(\hat{\mu}_i^2, \hat{\mu}_i^3)$	13.54	20.57	8.77	15.37	7.19	12.84	6.43	11.80	5.75	11.38
$(x_{i1}^2, x_{i1}^3, x_{i2}^2, x_{i2}^3)$	25.02	34.15	12.12	19.15	8.92	15.07	7.22	13.10	5.99	11.58
$(\hat{\eta}_i^2, \hat{\eta}_i^3)$	13.62	20.59	8.91	15.54	7.09	13.04	6.40	11.55	5.80	11.37
$(\hat{\mu}_i^2, \hat{\mu}_i^3, \hat{\mu}_i^4)$	18.84	26.33	10.55	17.57	8.30	14.31	7.02	12.19	5.31	10.36
$(x_{i1}^2, x_{i1}^3, x_{i1}^4, x_{i2}^2, x_{i2}^3, x_{i2}^4)$	39.58	48.80	17.03	25.47	11.37	18.20	8.68	15.08	6.46	12.42
$(\hat{\eta}_i^2, \hat{\eta}_i^3, \hat{\eta}_i^4)$	18.24	26.16	10.52	17.23	8.56	14.91	6.83	12.47	5.49	10.68

Table 5: Nonnull rejection rates under nonlinearity, $n = 20$.

added regressor(s)	$\alpha = 5\%$				$\alpha = 10\%$			
	δ				δ			
	1.2	1.4	1.6	1.8	1.2	1.4	1.6	1.8
$\hat{\mu}_t^2$	5.71	6.91	7.72	8.32	10.76	12.17	13.81	14.09
$\hat{\mu}_t^3$	7.77	17.94	34.40	54.21	14.47	28.41	48.42	68.72
$\hat{\mu}_t^4$	8.18	19.81	39.69	64.33	14.19	30.13	52.86	76.82
(x_{t1}^2, x_{t2}^2)	5.57	6.96	10.18	13.68	10.71	13.12	17.91	23.31
(x_{t1}^3, x_{t2}^3)	5.90	8.02	12.36	17.68	10.80	14.29	20.82	28.92
(x_{t1}^4, x_{t2}^4)	5.96	8.52	13.86	20.73	11.45	15.75	23.66	33.20
$\hat{\eta}_t^2$	8.09	19.44	39.81	65.57	14.05	29.97	53.25	77.50
$\hat{\eta}_t^3$	8.00	18.29	36.97	60.78	13.83	27.93	49.72	72.57
$\hat{\eta}_t^4$	7.84	16.80	33.51	54.72	13.61	26.08	45.49	67.67
$(\hat{\mu}_t^2, \hat{\mu}_t^3)$	7.40	15.73	31.70	53.84	13.22	25.04	44.25	67.69
$(x_{t1}^2, x_{t1}^3, x_{t2}^2, x_{t2}^3)$	5.83	7.40	11.28	16.41	10.91	14.06	19.54	27.34
$(\hat{\eta}_t^2, \hat{\eta}_t^3)$	7.50	16.01	32.06	54.39	13.24	25.08	44.63	67.96
$(\hat{\mu}_t^2, \hat{\mu}_t^3, \hat{\mu}_t^4)$	6.94	13.47	26.24	44.82	12.91	22.32	38.88	59.45
$(x_{t1}^2, x_{t1}^3, x_{t1}^4, x_{t2}^2, x_{t2}^3, x_{t2}^4)$	5.26	6.61	8.46	11.37	10.61	12.68	15.87	20.77
$(\hat{\eta}_t^2, \hat{\eta}_t^3, \hat{\eta}_t^4)$	7.34	15.09	31.03	55.49	13.68	23.97	44.48	69.05

Table 6: Nonnull rejection rates under nonlinearity, $n = 40$.

Added regressor(s)	$\alpha = 5\%$				$\alpha = 10\%$			
	δ				δ			
	1.2	1.4	1.6	1.8	1.2	1.4	1.6	1.8
$\hat{\mu}_t^2$	6.31	8.63	9.97	10.51	11.23	14.73	16.96	17.53
$\hat{\mu}_t^3$	10.45	30.71	59.87	84.02	18.01	43.30	72.98	91.04
$\hat{\mu}_t^4$	11.05	33.41	67.01	90.86	18.43	46.31	78.27	95.26
(x_{t1}^2, x_{t2}^2)	6.60	11.00	19.11	30.66	12.40	18.83	30.00	44.26
(x_{t1}^3, x_{t2}^3)	6.72	11.64	20.79	33.08	12.38	19.85	32.11	47.33
(x_{t1}^4, x_{t2}^4)	6.85	12.22	22.04	35.41	12.61	20.71	33.59	49.19
$\hat{\eta}_t^2$	11.21	33.34	68.82	92.14	18.14	45.87	78.90	95.96
$\hat{\eta}_t^3$	10.63	30.98	63.92	89.38	18.24	43.27	75.84	94.37
$\hat{\eta}_t^4$	9.97	28.24	58.47	84.66	17.30	39.88	70.84	91.73
$(\hat{\mu}_t^2, \hat{\mu}_t^3)$	9.06	25.93	56.71	85.23	16.23	37.67	69.18	91.74
$(x_{t1}^2, x_{t1}^3, x_{t2}^2, x_{t2}^3)$	6.04	9.20	14.19	22.17	11.63	16.37	23.95	34.49
$(\hat{\eta}_t^2, \hat{\eta}_t^3)$	9.04	25.86	56.69	85.19	16.25	37.58	69.29	91.89
$(\hat{\mu}_t^2, \hat{\mu}_t^3, \hat{\mu}_t^4)$	8.01	22.05	49.43	79.65	15.27	33.22	62.86	88.08
$(x_{t1}^2, x_{t1}^3, x_{t1}^4, x_{t2}^2, x_{t2}^3, x_{t2}^4)$	6.28	8.82	14.33	21.78	11.82	16.12	23.88	34.29
$(\hat{\eta}_t^2, \hat{\eta}_t^3, \hat{\eta}_t^4)$	9.52	27.02	60.35	89.03	16.56	38.56	71.99	94.10

Table 7: Nonnull rejection rates under nonlinearity, $n = 60$.

added regressor(s)	$\alpha = 5\%$ δ				$\alpha = 10\%$ δ			
	1.2	1.4	1.6	1.8	1.2	1.4	1.6	1.8
$\hat{\mu}_t^2$	8.21	14.95	20.61	26.96	14.45	23.21	30.92	37.77
$\hat{\mu}_t^3$	14.88	47.66	82.71	97.33	24.48	60.90	90.18	98.85
$\hat{\mu}_t^4$	16.32	53.58	89.13	99.23	25.77	66.29	94.03	99.74
(x_{t1}^2, x_{t2}^2)	7.45	17.32	34.61	56.55	14.16	27.43	47.91	70.25
(x_{t1}^3, x_{t2}^3)	7.47	18.59	37.95	61.42	14.48	29.25	51.71	74.46
(x_{t1}^4, x_{t2}^4)	7.77	19.71	40.51	65.10	14.88	30.66	54.29	77.21
$\hat{\eta}_t^2$	16.16	54.65	89.83	99.43	25.48	66.45	94.41	99.81
$\hat{\eta}_t^3$	15.48	51.47	86.86	98.88	24.13	63.18	92.65	99.48
$\hat{\eta}_t^4$	14.17	46.76	82.82	97.71	22.86	59.19	89.75	98.99
$(\hat{\mu}_t^2, \hat{\mu}_t^3)$	12.71	44.40	82.87	98.49	20.67	56.59	89.73	99.46
$(x_{t1}^2, x_{t1}^3, x_{t2}^2, x_{t2}^3)$	6.92	14.86	30.96	52.69	13.20	25.14	44.99	67.39
$(\hat{\eta}_t^2, \hat{\eta}_t^3)$	12.79	44.59	82.92	98.52	20.56	56.92	89.79	99.47
$(\hat{\mu}_t^2, \hat{\mu}_t^3, \hat{\mu}_t^4)$	10.91	38.35	76.80	97.36	18.26	50.81	85.61	98.83
$(x_{t1}^2, x_{t1}^3, x_{t1}^4, x_{t2}^2, x_{t2}^3, x_{t2}^4)$	6.93	14.26	28.64	49.63	12.97	23.50	41.37	63.99
$(\hat{\eta}_t^2, \hat{\eta}_t^3, \hat{\eta}_t^4)$	11.50	42.10	83.92	99.05	19.72	55.67	90.84	99.64

Table 8: Nonnull rejection rates under nonlinearity, $n = 100$.

added regressor(s)	$\alpha = 5\%$				$\alpha = 10\%$			
	δ				δ			
	1.2	1.4	1.6	1.8	1.2	1.4	1.6	1.8
$\hat{\mu}_t^2$	11.43	29.98	54.69	75.00	19.23	41.79	66.88	84.04
$\hat{\mu}_t^3$	17.26	54.92	87.42	98.13	26.43	67.08	93.32	99.25
$\hat{\mu}_t^4$	20.93	68.09	96.96	99.98	30.90	78.54	98.60	100.00
(x_{t1}^2, x_{t2}^2)	8.72	22.16	47.73	73.72	15.15	33.45	60.73	84.00
(x_{t1}^3, x_{t2}^3)	9.23	25.15	53.79	80.59	16.54	37.03	66.44	88.62
(x_{t1}^4, x_{t2}^4)	9.75	27.13	57.14	83.95	17.24	39.38	69.56	90.97
$\hat{\eta}_t^2$	21.62	70.14	97.75	99.99	30.99	79.40	98.90	100.00
$\hat{\eta}_t^3$	19.99	65.88	96.43	99.92	29.66	75.87	98.14	99.94
$\hat{\eta}_t^4$	17.97	60.48	94.09	99.80	27.89	71.24	96.91	99.92
$(\hat{\mu}_t^2, \hat{\mu}_t^3)$	16.52	60.06	95.15	99.94	26.07	71.43	97.49	99.99
$(x_{t1}^2, x_{t1}^3, x_{t2}^2, x_{t2}^3)$	7.69	20.55	46.83	75.81	14.61	32.05	60.67	85.72
$(\hat{\eta}_t^2, \hat{\eta}_t^3)$	16.67	60.80	95.35	99.96	26.25	71.65	97.56	100.00
$(\hat{\mu}_t^2, \hat{\mu}_t^3, \hat{\mu}_t^4)$	14.29	53.73	92.57	99.81	23.28	66.10	96.09	99.95
$(x_{t1}^2, x_{t1}^3, x_{t1}^4, x_{t2}^2, x_{t2}^3, x_{t2}^4)$	7.53	18.49	41.11	70.03	13.83	28.41	55.13	81.23
$(\hat{\eta}_t^2, \hat{\eta}_t^3, \hat{\eta}_t^4)$	16.41	62.27	96.85	100.00	26.31	73.95	98.76	100.00

the sample size ranges from $n = 20$ up to $n = 200$ and the nominal levels considered are $\alpha = 10\%, 5\%$.

The figures in Tables 9 to 11 lead to important conclusions. First, it is not an easy task to distinguish between logit and probit link functions. The power of the test in that setting is smaller than the corresponding powers under the other settings (complementary log-log and Cauchy links). For instance, when $n = 100$ and $\alpha = 5\%$, the best performing test has power around 30%. This was to be expected, since the logit and probit link functions are rather similar. One would need to have a quite large sample in order to be able to reliably distinguish between these two link functions. Here, the best performing test uses $\hat{\eta}_t^3$ as the single testing variable.

When the true link function is complementary log-log and the model is incorrectly specified using the logit link (Table 10), the powers are substantially larger than under probit data generation. The best test in this case is the one where the single testing variable is $\hat{\eta}_t^2$. When $n = 60$ and at the 5% nominal level, for example, the power of the test exceeds 90%. It is noteworthy that this test is substantially more powerful than the test of the exclusion of $\hat{\eta}_t^3$, which, under the same conditions, has power approximately equal to 1/3.

The misspecification associated with the Cauchy link (Table 11) leads to powers that are somewhat higher than those of the probit misspecification, but lower than the powers obtained when the complementary log-log link was the true link function. It is, however, possible to reliably identify the model misspecification when the sample size is not too small. For instance, when $n = 100$ and at the 5% nominal level the power of the best performing test, which uses $\hat{\eta}_t^3$ as the single testing variable, is in excess of 80%.

Finally, the results reported in Tables 9 to 11 indicate that the use of powers of the regressors as testing variables can lead to substantial power losses relative to the use of $\hat{\eta}_t^2$ or $\hat{\eta}_t^3$ as the single testing variable. The power loss in some cases is rather impressive. For instance, when the true link is complementary log-log (Table 10), the sample contains 60 observations and the test is performed at the 5% nominal level, the powers of the tests based on the testing variables $\hat{\eta}_t^2$ and (x_{t1}^2, x_{t2}^2) are 91.77% and 7.91%, respectively; the former is over 11 times larger than the latter.

5   Applications

In this section we shall revisit the two empirical applications briefly described in the Introduction and also consider a third application that uses a much larger sample size. At the outset, we turn to the first application. The main interest lies in modeling the proportion of crude oil converted to gasoline after distillation and fractionation. There are only ten sets of values of the three explanatory variables which correspond to ten different crudes subjected to experimentally controlled distillation conditions. The data were ordered according to the covariate that corresponds to the temperature

Table 9: Nonnull rejection rates with probit link.

added regressor(s)	n = 20		n = 40		n = 60		n = 100		n = 200	
	α		α		α		α		α	
	5%	10%	5%	10%	5%	10%	5%	10%	5%	10%
$\hat{\mu}_t^2$	8.99	16.38	11.09	19.50	12.85	20.61	22.36	33.31	34.42	46.46
$\hat{\mu}_t^3$	7.40	14.34	9.61	16.64	10.69	17.85	17.52	27.87	24.89	35.89
$\hat{\mu}_t^4$	6.67	12.88	8.41	15.01	9.33	16.18	14.49	24.42	19.26	28.79
(x_{t1}^2, x_{t2}^2)	5.98	11.92	5.72	11.40	5.38	10.30	6.41	12.42	8.64	15.57
(x_{t1}^3, x_{t2}^3)	5.54	11.28	5.25	10.97	5.07	10.33	5.76	11.30	7.38	13.79
(x_{t1}^4, x_{t2}^4)	5.44	10.78	5.14	10.56	4.73	10.19	5.18	10.80	6.79	12.72
$\hat{\eta}_t^2$	9.46	16.69	11.65	19.91	12.54	21.00	22.74	33.79	31.58	43.31
$\hat{\eta}_t^3$	12.36	19.92	15.48	23.58	17.22	26.12	30.87	42.72	48.51	61.40
$\hat{\eta}_t^4$	12.10	19.02	15.32	23.21	16.82	25.37	29.61	41.49	44.27	55.79
$(\hat{\mu}_t^2, \hat{\mu}_t^3)$	9.35	16.27	11.77	19.35	12.48	20.74	22.60	32.29	36.76	49.00
$(x_{t1}^2, x_{t1}^3, x_{t2}^2, x_{t2}^3)$	6.92	12.6	5.97	11.37	6.06	11.13	6.83	13.35	9.27	16.10
$(\hat{\eta}_t^2, \hat{\eta}_t^3)$	9.92	16.59	12.88	20.70	13.95	22.39	23.85	33.98	39.07	51.56
$(\hat{\mu}_t^2, \hat{\mu}_t^3, \hat{\mu}_t^4)$	8.43	15.09	10.67	18.37	11.78	19.24	20.59	30.49	33.71	44.98
$(x_{t1}^2, x_{t1}^3, x_{t1}^4, x_{t2}^2, x_{t2}^3, x_{t2}^4)$	6.68	12.48	6.12	11.64	6.01	11.89	6.88	13.51	8.59	15.27
$(\hat{\eta}_t^2, \hat{\eta}_t^3, \hat{\eta}_t^4)$	8.94	15.42	10.92	18.84	12.28	20.20	20.67	31.51	33.73	45.92

Table 10: Nonnull rejection rates with complementary log-log link.

added regressor(s)	n = 20		n = 40		n = 60		n = 100		n = 200	
	α		α		α		α		α	
	5%	10%	5%	10%	5%	10%	5%	10%	5%	10%
$\hat{\mu}_t^2$	35.31	49.89	56.72	70.65	78.03	87.21	95.28	97.71	99.99	100.00
$\hat{\mu}_t^3$	48.11	61.46	71.15	81.74	89.82	94.30	98.51	99.37	100.00	100.00
$\hat{\mu}_t^4$	52.29	65.30	74.37	84.20	92.05	95.64	98.91	99.59	100.00	100.00
(x_{t1}^2, x_{t2}^2)	7.16	13.37	10.44	18.15	7.91	15.13	18.21	28.67	72.43	83.59
(x_{t1}^3, x_{t2}^3)	5.78	11.44	10.11	17.47	6.75	13.37	15.75	26.25	65.84	78.47
(x_{t1}^4, x_{t2}^4)	4.71	9.68	9.30	16.52	5.89	11.95	14.49	23.64	59.10	71.89
$\hat{\eta}_t^2$	50.75	63.93	72.82	83.39	91.77	95.49	98.82	99.54	100.00	100.00
$\hat{\eta}_t^3$	4.40	9.18	9.65	17.26	33.04	45.64	21.68	32.51	98.13	99.24
$\hat{\eta}_t^4$	46.92	60.89	67.08	77.75	89.06	93.69	97.95	99.07	100.00	100.00
$(\hat{\mu}_t^2, \hat{\mu}_t^3)$	39.53	53.33	63.78	75.15	85.58	92.09	97.65	98.94	100.00	100.00
$(x_{t1}^2, x_{t1}^3, x_{t2}^2, x_{t2}^3)$	6.84	13.65	6.34	12.50	7.30	14.52	21.51	32.51	60.28	73.34
$(\hat{\eta}_t^2, \hat{\eta}_t^3)$	41.18	53.65	65.14	76.22	86.69	92.61	97.88	99.02	100.00	100.00
$(\hat{\mu}_t^2, \hat{\mu}_t^3, \hat{\mu}_t^4)$	34.80	46.93	56.91	69.44	81.23	89.03	96.47	98.34	100.00	100.00
$(x_{t1}^2, x_{t1}^3, x_{t1}^4, x_{t2}^2, x_{t2}^3, x_{t2}^4)$	5.16	10.61	6.76	13.01	7.52	14.80	17.84	29.04	54.22	68.01
$(\hat{\eta}_t^2, \hat{\eta}_t^3, \hat{\eta}_t^4)$	35.03	47.46	57.90	69.94	81.65	88.94	96.49	98.41	100.00	100.00

Table 11: Nonnull rejection rates with Cauchy link.

added regressor(s)	n = 20		n = 40		n = 60		n = 100		n = 200	
	α		α		α		α		α	
	5%	10%	5%	10%	5%	10%	5%	10%	5%	10%
$\hat{\mu}_t^2$	21.85	34.02	33.48	45.79	41.00	54.22	75.46	84.23	89.87	94.98
$\hat{\mu}_t^3$	19.05	29.02	28.38	40.52	33.67	46.16	67.12	78.41	81.06	88.62
$\hat{\mu}_t^4$	16.63	25.03	24.37	35.86	28.91	39.73	58.23	72.11	69.51	80.48
(x_{t1}^2, x_{t2}^2)	9.53	16.63	9.19	16.66	7.74	13.95	13.30	22.52	26.57	38.23
(x_{t1}^3, x_{t2}^3)	7.95	14.73	8.38	15.40	6.88	12.43	10.48	18.18	20.34	30.85
(x_{t1}^4, x_{t2}^4)	6.79	13.49	7.40	14.28	5.93	11.71	8.73	15.72	16.03	26.08
$\hat{\eta}_t^2$	20.70	32.50	32.37	44.15	37.82	51.06	72.14	83.11	86.09	92.36
$\hat{\eta}_t^3$	27.12	39.95	37.96	52.22	50.03	62.69	82.10	89.43	97.21	98.76
$\hat{\eta}_t^4$	25.53	38.50	34.81	48.47	45.59	58.71	78.39	87.24	94.05	97.08
$(\hat{\mu}_t^2, \hat{\mu}_t^3)$	19.88	31.72	29.62	42.42	39.44	52.54	73.08	82.99	95.02	97.49
$(x_{t1}^2, x_{t1}^3, x_{t2}^2, x_{t2}^3)$	9.29	17.49	7.71	14.82	8.13	14.61	16.06	27.10	28.48	39.66
$(\hat{\eta}_t^2, \hat{\eta}_t^3)$	19.65	31.30	28.61	41.42	39.00	52.54	71.27	81.97	94.40	97.36
$(\hat{\mu}_t^2, \hat{\mu}_t^3, \hat{\mu}_t^4)$	15.83	26.20	24.13	36.44	33.21	46.28	65.65	77.04	92.34	95.91
$(x_{t1}^2, x_{t1}^3, x_{t1}^4, x_{t2}^2, x_{t2}^3, x_{t2}^4)$	8.08	14.69	7.02	14.04	8.44	16.24	14.31	24.27	23.82	35.83
$(\hat{\eta}_t^2, \hat{\eta}_t^3, \hat{\eta}_t^4)$	15.53	26.00	24.34	36.81	33.79	46.15	65.87	77.43	92.18	95.89

at which 10% of the crude oil vaporizes. This variable assumes ten different values which are used to define the ten batches of crude oil. The model specification uses an intercept ($x_1 = 1$), nine dummy variables for the first nine batches of crude oil ($x_2, \ldots, x_{10}$) and the covariate that measures the temperature (degrees F) at which all the gasoline is vaporized (x_{11}). There are $n = 32$ observations. The maximum likelihood point estimates obtained using the logit link function are presented, along with the corresponding asymptotic standard errors and p-values, in Table 12.

Table 12: Parameter estimates using Prater's gasoline data.

parameter	estimate	standard error	p-value
β_1	−6.15957	0.18232	0.00000
β_2	1.72773	0.10123	0.00000
β_3	1.32260	0.11790	0.00000
β_4	1.57231	0.11610	0.00000
β_5	1.05971	0.10236	0.00000
β_6	1.13375	0.10352	0.00000
β_7	1.04016	0.10604	0.00000
β_8	0.54369	0.10913	0.00000
β_9	0.49590	0.10893	0.00001
β_{10}	0.38579	0.11859	0.00114
β_{11}	0.01097	0.00041	0.00000
ϕ	440.27838	110.02562	—

We aim at providing the following question with an answer: Is the model used by Ferrari & Cribari-Neto (2004), whose estimates are given in Table 12, correctly specified? We apply to the test proposed in Section 3 to the model at hand. We have used the following testing variables: $\hat{\eta}_t^2$, $\hat{\eta}_t^3$, $\hat{\eta}_t^4$, $\hat{\mu}_t^2$, $\hat{\mu}_t^3$ and $\hat{\mu}_t^4$. The respective score test statistics are 16.401, 16.303, 15.537, 14.711, 13.065 and 11.682. Thus, all tests reject the null hypothesis of correct model specification even at very small nominal levels. Therefore, there is strong evidence that the model is *not* correctly specified.

We proceeded by estimating beta regressions using four different link functions, namely: probit, Cauchy, complementary log-log and log-log. The same testing variables were used. The null hypothesis of no misspecification was rejected at the 10% nominal level for all link functions, except for the log-log link function (test statistics, respectively, equal to 1.6277, 1.2394, 1.1846, 1.5269, 1.659 and 1.7458).

It is noteworthy that the pseudo-R^2 of the log-log beta regression model equals 0.99 whereas that of the logit model equals 0.96. Figure 2 plots $\hat{\mu}_t$ (fitted values) against y_t (observed responses) for the two models. It is clear that the log-log link function yields a model whose fit better accommodates the two (lower and upper) extreme observations, and also some observations in the middle portion of the data. We take that as further evidence in favor of the log-log model.

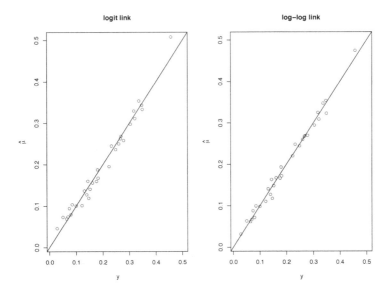

Figure 2: Fitted values plotted against observed responses: logit and log-log models.

The conclusion we reach is that the model specification used by Ferrari & Cribari-Neto (2004) is not correctly specified. In particular, the link function (logit) is mis-specified. Correct model specification requires the use of the log-log link function $g(\mu) = -\log\{-\log(\mu)\}$.

We now move to the second application. As noted in the Introduction, the variable of interest (y) are reading accuracy indices of 44 Australian children. The covariates are: dyslexia versus non-dyslexia status (x_2), nonverbal IQ converted to z-scores (x_3) and an interaction variable (x_4). The participants (19 dyslexics and 25 controls) were students from primary schools in the Australian Capital Territory. The children ages range from eight years five months to twelve years three months. The regressor x_2 is a dummy variable, which equals 1 if the child is dyslexic and -1 otherwise. The observed scores were linearly transformed from their original scale to the open unit interval $(0, 1)$; see Smithson & Verkuilen (2006). The mean accuracy was 0.900 for non-dyslexic readers and 0.606 for dyslexic children. The scores ranged from 0.459 to 0.990, and averaged 0.773. Espinheira et al. (2008) and Smithson & Verkuilen (2006) found evidence of nonconstant dispersion. Their beta regression model was specified as $g(\mu_t) = \beta_1 + \beta_2 x_{t2} + \beta_3 x_{t3} + \beta_4 x_{t4}$ and $\log(\phi_t) = -(\gamma_1 + \gamma_2 x_{t2} + \gamma_3 x_{t3})$. The maximum likelihood estimates of $\beta_1, \ldots, \beta_4$ ($\gamma_1, \ldots, \gamma_3$) are 1.1232, -0.7416, 0.4864 and -0.5813 (-3.3044, -1.7466 and -1.2291), respectively. They used the logit link function for the mean response. What can be said about such specification? We carried the score RESET test using $\hat{\eta}_t^2$ as testing variable for five different link functions, namely: logit, probit, complementary log-log, log-log and Cauchy. The respective p-values were 0.3150, 0.5016, 0.7447, 02685 and 0.0534. We note that the only substantial evidence of incorrect model specification occurs when the Cauchy link is used (at the 10% nominal level). We have also estimated the model with no interaction effect, i.e., $g(\mu_t) = \beta_1 + \beta_2 x_{t2} + \beta_3 x_{t3}$, where $g(\cdot)$ is the logit link. The test p-value was 0.0583, i.e., the test suggests that the model with no interaction between dyslexia and IQ is not correctly specified (at the 10% nominal level). In conclusion, there is no evidence that the beta regression model used by Smithson & Verkuilen (2006) is incorrectly specified.

Our third application uses a much larger sample size: nearly five thousand observations. Our response variable is the difference between the proportions of valid votes for the candidate Luiz Inácio Lula da Silva in the runoff (i.e., second round) elections of 2006 and 2002 for the presidency of Brazil.[2] He won both elections: he was elected president of Brazil in 2002 and reelected for the same post in 2006. It is noteworthy that shortly after taking office in 2002 president Lula unified existing several assistance programs into a program called 'Bolsa-Família' and greatly increased its budget over the years. Under the Bolsa-Família program poor families receive monthly cash payments from the Federal Government. It has thus being claimed that the increased spending on cash payments to poor families between 2002 and 2006

---

[2]The variate was normalized in order to assume values in the standard unit interval.

impacted the outcome of the 2006 presidential election in Brazil, yielding a large number of votes for the candidate that was running for reelection that he would not otherwise receive; see, e.g., Zucco (2008).

We have estimated a varying dispersion beta regression model in order to determine whether the increased spending on the Bolsa-Família program between 2002 and 2006 played a noticeable role in the 2006 election.[3] We consider several regressors, namely: (i) x_2: difference between the total spending on assistance programs for the poor (Bolsa-Família and other similar programs) in 2006 and 2002 (adjusted for inflation), (ii) x_3: average income in the county, (iii) x_4: dummy that equals one if the governor belongs to the same political party as Lula and zero otherwise, (iv) x_5: dummy that equals one if the mayor belongs to the same political party as Lula and zero otherwise; (v) x_6: dummy variable that equals one if the county had more than 50 thousand inhabitants in 2006 and zero otherwise, (vi) x_7: proportion of all county inhabitants that live in urban areas, (viii) x_8: proportion of people over 15 years old that are illiterate. The source of the data is Zucco (2008) and Instituto Brasileiro de Geografia e Estatística (the Brazilian census bureau). All regressors were included in the mean submodel (i.e., the regression model for μ) and the following variables entered the submodel for ϕ (nonconstant precision): x_2 and x_7. The link functions used for the mean and precision submodels were loglog and square root, respectively. The mean effect was modeled as $\mathrm{loglog}(\mu_t) = \beta_1 + \beta_2 x_{t2} + \beta_3 x_{t3} + \beta_4 x_{t4} + \beta_5 x_{t5} + \beta_6 x_{t6} + \beta_7 x_{t7} + \beta_8 x_8$ and the precision as $\sqrt{\phi_t} = \gamma_1 + \gamma_2 x_{t2} + \gamma_3 x_{t7}$. Parameter estimation was carried by maximum likelihood using data from 4,864 Brazilian counties. To save space, we do not present the point estimates. We note, nonetheless, that all covariates are statistically significant at the 1% nominal level and that the pseudo-R^2 equals 0.66. The model thus yields a good fit. The estimated coefficient of x_2 is positive (0.001922) and highly significant, thus implying that increases in spending on the Bolsa-Família program positively impacted the proportion of valid votes for Lula in the 2006 election.

Since the sample size is large, we carry out the RESET misspecification test using the likelihood ratio test and use 1% as the benchmark nominal level. As before, we use $\hat{\eta}_t^2$ as testing variable. The test statistic equals 5.5281 (p-value > 0.01), and we do not reject the null hypothesis that the model is correctly specified. The same conclusion is reached when the precision link function is changed to log: test statistic equal to 6.5281 and p-value > 0.01. That is, we find no evidence of model misspecification. Likewise we find no evidence against the model that uses the Cauchy and squared root links, the test statistic value being 4.5035 (p-value > 0.01); when the precision link is log, however, the test statistic equals 8.5904 (p-value < 0.01), and

---

[3]Zucco (2008) estimated linear regressions using a different response variable. He also used different regressors. In particular, he considered the proportion of families that receive the Bolsa-Família cash payment in each county whereas we use the total spending per county, which also reflects real (i.e., after adjusting for inflation) increases in the amount each family receives.

the model is rejected. When the mean submodel link function is logit, the test statistic values are 18.602 (square root link for the precision submodel) and 19.728 (log link for the precision submodel), the p-values being very close to zero. The corresponding test statistic values for the probit and complementary log-log links are: (i) 20.627 and 21.857, and (ii) 38.440 and 40.170. We thus strongly reject the null hypothesis of correct model specification.

6   Concluding remarks and directions for future research

The beta regression model is useful for modeling responses that assume values on the standard unit interval, such as rates and proportions. The model allows practitioners to condition the mean behavior of the variable of interest on a set of explanatory variables. In this paper, we have proposed a misspecification test for beta regressions. The null hypothesis under test is that the beta regression is correctly specified, whereas under the alternative hypothesis the model specification is in error. The proposed test is an extension of the RESET test of Ramsey (1969), which was devised for the classic linear regression model. It follows from augmenting the model with a set of testing variables whose exclusion can be tested using the likelihood ratio, Rao's score or the Wald test. The numerical evidence favors the score test. An advantage of the score test is that it only requires estimation of the null model which is, under the null hypothesis, corectly specified. Additionally, the Monte Carlo evidence showed that the proposed test can be useful when there are neglected nonlinearities and when the link function is not correctly specified. We have also presented three empirical applications (two that uses small samples and one that is based on a large number of observations).

We encourage practitioners to use the RESET test when estimating beta regressions. In particular, we recommend the use of the score implementation of the test along with the squared or cubed fitted linear predictor as testing variable.

Our analysis focused on the standard beta regression model, i.e., we considered the fully parametric model as proposed by Ferrari & Cribari-Neto (2004). An important direction for future research is the evaluation of the RESET specification test in a wider class of models, such as the class of Generalized Additive Models for Location, Scale and Shape (GAMLSS); see Rigby & Stasinopoulos (2005). It is noteworthy that the GAMLSS class allows investigators to include nonparametric components in the mean response and dispersion specifications. Future research should also consider the development of tests for testing a beta regression against smooth alternatives, along the lines of Kauerman & Tutz (2001).

The specification test considered in this paper can be used, as explained and illustrated, to evaluate the adequacy of a particular link function. Our approach was fully frequentist. A Bayesian formulation of the beta regression model was developed by Branscum et al. (2007) and a Bayesian approach for link function selection in the class of generalized linear models was proposed by Czado & Raftery (2006). The evaluation of some variant of their approach in the class of beta regressions is also an important direction for future research.

References

Branscum, A. J., Johnson, W. O. & Thurmond, M. C. (2007), 'Bayesian beta regression: applications to household expenditure data and genetic distance between foot-and-mouth disease viruses', *Australian and New Zealand Journal of Statistics* **49**, 287–301.

Cribari-Neto, F. & Vasconcellos, K. L. (2002), 'Nearly unbiased maximum likelihood estimation for the beta distribution', *Journal of Statistical Computation and Simulation* **72**, 107–118.

Cribari-Neto, F. & Zeileis, A. (2010), 'Beta regression in R', *Journal of Statistical Software* **34**(2), 1–24.

Czado, C. & Raftery, A. E. (2006), 'Choosing the link function and accounting for link uncertainty in generalized linear models using bayes factors', *Statistical Papers* **47**, 419–442.

Daniel, C. & Wood, F. S. (1971), *Fitting Equations to Data*, Wiley, New York.

Doornik, J. A. (2001), *Ox: an Object-oriented Matrix Programming Language*, Timberlake Consultants, London.

Espinheira, P. L., Ferrari, S. L. P. & Cribari-Neto, F. (2008), 'Influence diagnostics in beta regression', *Computational Statistics and Data Analysis* pp. 4417–4431.

Ferrari, S. L. P. & Cribari-Neto, F. (2004), 'Beta regression for modelling rates and proportions', *Journal of Applied Statistics* **31**, 799–815.

Kauerman, G. & Tutz, G. (2001), 'Testing generalized linear and semiparametric models against smooth alternatives', *Journal of the Royal Statistical Society B* **63**, 147–166.

Kieschnick, R. & McCullough, B. (2003), 'Regression analysis of variates observed on (0,1): percentages, proportions, and fractions', *Statistical Modelling* **3**, 193–213.

McCullagh, P. & Nelder, J. A. (1989), *Generalized Linear Models*, 2nd ed., Chapmann and Hall, London.

Ospina, R., Cribari-Neto, F. & Vasconcellos, K. L. P. (2006), 'Improved point and interval estimation for a beta regression model', *Computational Statistics and Data Analysis* **51**, 960–981. Errata: vol. 55, p. 2445, 2011.

Paolino, P. (2001), 'Maximum likelihood estimation of models with beta-distributed dependent variables', *Political Analysis* **9**, 325–346.

Prater, N. H. (1956), 'Estimate gasoline yields from crudes', *Petroleum Refiner* **35**, 236–238.

Press, W. H., Teukolsky, S. A., Vetterling, W. T. & Flannery, B. P. (1992), *Numerical Recipes in C: the Art of Scientific Computing*, 2nd ed., Cambridge University Press, New York.

Ramsey, J. B. (1969), 'Tests for specification errors in classical linear least squares regression analysis', *Journal of the Royal Statistical Society B* **31**, 350–371.

Rigby, B. & Stasinopoulos, M. (2005), 'Generalized additive models for location, scale and shape', *Applied Statistics* **54**, 507–554.

Simas, A. B., Barreto-Souza, W. & Rocha, A. V. (2010), 'Improved estimators for a general class of beta regression models', *Computational Statistics and Data Analysis* **54**, 348–366.

Smithson, M. & Verkuilen, J. (2006), 'A better lemon-squeezer? maximum likelihood regression with beta-distribuited dependent variables', *Psychological Methods* **11**, 54–71.

Vasconcellos, K. L. P. & Cribari-Neto, F. (2005), 'Improved maximum likelihood estimation in a new class of beta regression models', *Brazilian Journal of Probability and Statistics* **19**, 13–31.

Zucco, C. (2008), 'The president's 'new' constituency: Lula and the pragmatic vote in Brazil's 2006 presidential election', *Journal of Latin American Studies* **40**, 29–39.

Considering correlation properties on statistical simulation of clutter[‡]

Ana Georgina Flesia[*]   María Magdalena Lucini[†]
Dario Javier Perez[*]

[*] Facultad de Matemática Astronomía y Física, y CIEM-Conicet
Universidad Nacional de Córdoba,
Ing. Medina Allende s/n
5000 Córdoba, Argentina,
Fax: +54-351-4334054

flesia,djp0109@famaf.unc.edu.ar

[†] Universidad Nacional de Nordeste y Conicet
Facultad de Ciencias Exactas, Naturales y Agrimensura,
Av. Libertad 5450 - Campus "Deodoro Roca"
(3400) Corrientes
Tel: +54 (3783) 473931/473932

lucini@exa.unne.edu.ar

1   Introduction

An imaging radar is a system for earth observation based on an emitting and receiving device that operates in the range of microwaves. The system sends a pulse of electromagnetic energy, the targets reacts to this stimulus and, eventually, part of this energy is returned to the system. This return signal, if available, is processed to infer about the properties of the target.

Imaging radar systems constitute a major advance in remote sensing, since they allow the obtainment of dielectric properties of targets independently of the availability of natural illumination (they carry their own source of energy) and of the weather conditions (microwaves are unaffected, to a great extent, by clouds, fog, rain, smog etc). Besides these desirable properties, the bandwidth of the signal employed is able to penetrate canopy and other masses.

[‡]This work was financially supported by grants from SeCyT-UNC, Argentina. AGF, MML are career members of CONICET.

The term *synthetic* refers to the fact that larger antennas and, thus, greater resolutions, are obtained with processing techniques. These characteristics allow the use of synthetic aperture radar (SAR) systems for continuous earth monitoring.

The statistical properties here presented are common to every image generated with coherent illumination, as is the case of ultrasound, laser and sonar. The relevant information present in these images is concentrated in the mean cross section. This quantity is sensitive to many parameters that characterize the target, as dielectric constant and surface roughness, among other. Each individual cell in the image (pixel) has this information, but it is corrupted by the *speckle* noise, which is due to interference phenomena in the reflected signal.

The demand of exhaustive clutter measurement in all scenarios would be alleviated if plausible data could be obtained by computer simulation. The adoption of correlated clutter model is significant since it is the correlation effects within the clutter which often dominate system performance.

The purpose of this work is to impulse further the use of correlation in simulation studies by designing a package in the free language R to simulate data with correlated properties, making it available for download following the Reproducible Research Paradigm. The simulation techniques we discuss here have been introduced in Bustos et al. (2001) for the $\mathcal{K}$ distribution, and Bustos et al. (2009) for the $\mathcal{G}^0$ distribution. The examples reported by the authors were implemented using proprietary software, the IDL 5.1 development platform, with a set of auxiliary Fortran routines. Moreover Bustos & Frery (2006) showed that IDL may present numerical instabilities, while Almiron et al. (2010) conducted an analysis of platforms showing that R (freely available at http://www.r-project.org) has excellent numerical performance. We consider thus an important contribution to the community of image processing researchers the availability of R simulation routines that introduce correlation structure in data. Code have been tested in linux and windows platforms.

We give an example of the code versatility showing that accuracy of non-parametric techniques change when correlated data is classified, compared with classification of uncorrelated data simulated with the same parameters. Spatial correlation introduced in the clustering paradigm as a priori information in the map of classes increments accuracy of common nonparametric methods as k-means and ISODATA, when Iterated Conditional Modes (ICM) is used to estimate the final map of classes. Classical methods over correlated data gives high discrimination without the need of a priori information. All code, examples plus simulation algorithms, is available for download from AGF's Reproducible Research website.

In the following section we will review the statistical properties of SAR data. In section 3 we we review some of the techniques to simulate correlation properties, and in section 4 we introduce the R package for simulation within an example of accuracy of nonparametric classification.

2   The multiplicative model and the speckle noise

The multiplicative model has been widely used in the modeling, processing, and analysis of synthetic aperture radar images. This model states that, under certain conditions the return results from the product between the speckle noise and the terrain backscatter, see   Mejail et al. (2001) and references therein.

Based upon this model, we assume that the observed value in each pixel within this kind of images is the outcome of the product of two independent two dimensional random processes: one X modeling the terrain backscatter, and other Y modeling the speckle noise. The former is many times considered real and positive, while the latter could be complex (if the considered image is in the complex format) or positive and real (intensity and amplitude formats). Therefore, the observed value is the outcome of the random process defined by the product

$$Z_{(s_1,s_2)} = X_{(s_1,s_2)} Y_{(s_1,s_2)} \qquad \forall\, (s_1,s_2) \in Z^2, \qquad (1)$$

where (s_1, s_2) denotes the spatial position of the pixel. We will say that the process Z_I is the intensity return process if $Z_I = |Z|^2$, and Z_A is the amplitude return process if $Z_A = |Z|$.

The complex format has been used as a flexible tool for the statistical modeling of SAR data. However, in several cases, complex data are not available or exists computational limitations imposed by the imaging system that not allow us to work with them. As a consequence, intensity format and amplitude format are more frequently considered in the literature.

In many cases, it is easier to derive the statistical properties of the intensity data rather than amplitude data. For instance, the intensity speckle noise modeled as the sum of independent and exponentially distributed random variables has well know distribution, the Gamma distribution, but this is not the case for amplitude speckle noise, since the convolution of Rayleigh distributions has not closed from, Frery et al. (1997).

Multilook data results from taking the average over n independent samples $Z_r(s_1, s_2) = X_{(s_1,s_2)} Y_r(s_1, s_2)$ $1 \leq r \leq n$, this is

$$\hat{Z}_n(s_1, s_2) = \frac{1}{n} \sum_{r=1}^{n} Z_r(s_1, s_2) = X_{(s_1,s_2)} \hat{Y}_n(s_1, s_2), \tag{2}$$

where $\hat{Y}_n$ is the n-look average speckle, since X (the target9 does not vary from image to image.

Following the description that Frery et al. (1997) made about the appropriated distributions for this model, complex speckle is assumed to have a bivariate normal distribution, with independent identically distributed components having zero mean and variance $1/2$. These marginal distributions are denoted here as $N(0, 1/2)$, therefore, $Y_{C,(s_1,s_2)} = (Re(Y_{(s_1,s_2)}), Im(, Y_{(s_1, s_2)})) \sim N^2(\mathbf{0}, \mathbf{1/2})$ denotes the distribution of a pair.

Multilook intensity speckle results from taking the average over n independent samples of $Y_{I,(s_1,s_2)} = |Y_{C,(s_1,s_2)}|^2$ leading, thus, to a Gamma distribution denoted here as $Y_{I,(s_1,s_2)} \sim \Gamma(n, n)$ and characterized by the density

$$f_{Y_I}(y) = \frac{n^n}{\Gamma(n)} y^{n-1} e^{-ny} \qquad y > 0, n > 0. \tag{3}$$

The multilook amplitude speckle can be obtained as the square root of multilook intensity speckle, leading to a square root of Gamma distribution denoted by $Y_A(s_1, s_2) \sim \Gamma^{1/2}(n, n)$ and characterized by the density

$$f_{Y_A}(y) = \frac{2n^n}{\Gamma(n)} y^{2n-1} e^{-ny^2} \qquad y > 0, n > 0. \tag{4}$$

Several distributions could be used for the backscatter, aiming at the modeling of different types of classes and their characteristic degrees of homogeneity. For instance, for some sensor parameters (wavelength, angle of incidence, polarization, etc), pasture is more homogeneous than forest, which, in turn, is more homogeneous than urban areas. Such distributions are, in the case of intensity backscatter,

a) a constant, β^2, when the target area is homogeneous,

b) when the region is non homogeneous, the Gamma distribution, denoted by $X_{I,(s_1,s_2)} \sim \Gamma(\alpha, \lambda)$, and characterized by the density

$$f_{X_I}(x) = \frac{\lambda^\alpha}{\Gamma(\alpha)} x^{\alpha-1} e^{-\lambda x}, \qquad x > 0, \alpha > 0, \lambda > 0. \tag{5}$$

c) For extremely heterogeneous regions, the reciprocal of a Gamma distribution, denoted by $X_I(s_1, s_2) \sim \Gamma^{-1}(\alpha, \gamma)$ and characterized by the density

$$f_{X_I}(x) = \frac{1}{\Gamma(\alpha)\gamma^\alpha} x^{\alpha-1} e^{-\frac{\gamma}{x}}, \qquad x > 0, \ -\alpha > 0, \ \gamma > 0. \qquad (6)$$

In the case of the amplitude backscatter X_A, the formula $X_A = \sqrt{X_I}$ leads to the following distributions,

a) a constant, β, when the target area is homogeneous,

b) when the region is non homogeneous, the square root of Gamma distribution, denoted by $X_A(s_1, s_2) \sim \Gamma^{1/2}(\alpha, \lambda)$, and characterized by the density

$$f_{X_A}(x) = \frac{2\lambda^\alpha}{\Gamma(\alpha)} x^{2n-1} e^{-\lambda x^2}, \qquad x > 0, \ \alpha > 0, \ \lambda > 0, \qquad (7)$$

c) For extremely heterogeneous regions, the reciprocal of a square root of Gamma distribution, denoted by $X_A(s_1, s_2) \sim \Gamma^{-1/2}(\alpha, \gamma)$, and characterized by the density

$$f_{X_A}(x) = \frac{2}{\Gamma(\alpha)\gamma^\alpha} x^{2\alpha-1} e^{-\frac{\gamma}{x^2}}, \qquad x > 0, \ -\alpha > 0, \ \gamma > 0. \qquad (8)$$

The distribution of the return arises from the product $Z = X.Y$; its density is the result of the convolution of the densities of the backscatter and speckle noise. For instance, in the homogeneous case, we consider X_I a constant β^2 and the multilook intensity speckle $X_I \sim \Gamma(n, n)$, then the return Z_I can be modeled by a Gamma distribution, denoted by $Z_I(s_1, s_2) \sim \Gamma(n, n/\beta^2)$. The following list summarize the distributions for the intensity return

a) a Gamma distribution, denoted by $Z_I(s_1, s_2) \sim \Gamma(n, n/\beta^2)$, when the target area is homogeneous,

b) for regions with heterogeneous texture, Z_I is said obey the $\mathcal{K}_I$ distribution, situation here denoted as $Z_I(s_1, s_2) \sim \mathcal{K}_I(\alpha, \gamma, n)$, if its density is given by

$$f_{Z_I}(z) = \frac{2\left(\sqrt{\lambda n}\right)^{n+\alpha}}{\Gamma(\alpha)\Gamma(n)} z^{\frac{n+\alpha}{2}-1} K_{n-\alpha}\left(2\sqrt{\lambda n z}\right) \qquad z > 0, \ \alpha > 0, \ \lambda > 0, \ n > 0. \qquad (9)$$

c) the $\mathcal{G}_I^0$ distribution for extremely heterogeneous areas. This distribution, denoted by $Z_I(s_1, s_2) \sim \mathcal{G}_I^0(\alpha, \gamma, n)$, is characterized by the density

$$f_{Z_I}(z) = \frac{n^n \Gamma(n - \alpha) z^{n-1}}{\gamma^\alpha \Gamma(-\alpha)(\gamma + nz)^{n-\alpha}}, \qquad z > 0, \ -\alpha > 0, \ \gamma > 0, \ n > 0. \quad (10)$$

The following list summarize the distributions for the amplitude return Z_A

a) square root of Gamma distribution denoted by $Z_A(s_1, s_2) \sim\sim \Gamma^{-1/2}(n, n/\beta)$, usada para modelar áreas homogéneas,

b) for regions with heterogeneous texture, Z_A is said obey the $\mathcal{K}_A$ distribution, denoted as $Z_A(s_1, s_2) \sim \mathcal{K}_A(\alpha, \gamma, n)$, with density given by

$$f_{Z_A}(z) = \frac{4 (\lambda n)^{\frac{n+\alpha}{2}}}{\Gamma(\alpha)\Gamma(n)} z^{n-\alpha+1} K_{n-\alpha}\left(2z\sqrt{\lambda n}\right) \qquad z > 0, \ \alpha > 0, \ \lambda > 0, \ n > 0. \quad (11)$$

where K_ν is the modified Bessel function of the third kind and order ν.

c) the $\mathcal{G}_A^0$ distribution for extremely heterogeneous areas, denoted by $Z_I(s_1, s_2) \sim \mathcal{G}_I^0(\alpha, \gamma, n)$, is characterized by the density

$$f_{Z_I}(z) = \frac{2n^n \Gamma(n - \alpha) z^{2n-1}}{\Gamma(n)\gamma^\alpha \Gamma(-\alpha)(\gamma + nz^2)^{n-\alpha}}, \qquad z > 0, \ -\alpha > 0, \ \gamma > 0, \ n > 0. \quad (12)$$

In Table 1 we summarize the above distributions for backscatter, return and noise.

3   Simulation perspective: Inverse Transform Method

In formulating a stochastic model to describe a real phenomenon, there is always a compromise between choosing a model that is a realistic replica of the actual situation and choosing one whose mathematical analysis is

Process	Intensity		
	homog.	heter.	extre. heter.
Backscatter X	β^2	$\Gamma(\alpha,\lambda)$	$\Gamma^{-1}(\alpha,\gamma)$
Noise Y	$\Gamma(n,n)$		
Return Z	$\Gamma(n,n/\beta^2)$	$\mathcal{K}_I(\alpha,\lambda,n)$	$\mathcal{G}_I^0(\alpha,\gamma,n)$
Process	Amplitud		
	homog.	heter.	extre. heter.
Backscatter X	β	$\Gamma^{1/2}(\alpha,\lambda)$	$\Gamma^{-1/2}(\alpha,\gamma)$
Noise Y	$\Gamma^{1/2}(n,n)$		
Return Z	$\Gamma^{1/2}(n,n/\beta)$	$\mathcal{K}_A(\alpha,\lambda,n)$	$\mathcal{G}_A^0(\alpha,\gamma,n)$

Table 1:   Table of distributions for intensity and amplitude format n-look images

tractable. That is, there is no payoff in choosing a model faithfully conformed to the phenomenon under study if it were not possible to mathematically analyze the model. However, the relatively recent advance of fast and inexpensive computational power has opened up another approach-namely try to model the phenomenon as faithfully as possible and then to rely on a simulation study to analyze it.

Table 1 shows the full extent of the problem of simulation of textures under the statistical model of SAR images. In general, there is a classical approach to the problem of generating outcomes of correlated vectors, called the inverse transform method, Ross (2012). A modification of such method, summarized in the following three steps, was proposed by Flesia (1999) in her PhD thesis, and particularized for the constructions of correlated *Gamma* vectors:

1. generating independent outcomes from a convenient distribution;

2. introducing correlation in these data;

3. transforming the correlated observations into the desired marginal properties.

The transformation that guarantees this is obtained from the cumulative distribution functions of the data obtained in step 2 and that of the desired distributions. The reader is invited to recall that if U is a random variable with cumulative distribution function F_U then $F_U(U)$ obeys a $\mathcal{U}(0,1)$ law and, reciprocally, if V obeys a $\mathcal{U}(0,1)$ distribution then $F_U^{-1}(V)$ is F_U distributed, Ross (2012). If the expressions for resulting correlations after the transformation are available beforehand it is possible, in principle, to perform step 2 such that, after the transformation, the desired correlation structure is obtained.

In principle, there are no restrictions on the possible order parameters values that can be obtained by this method, but numerical issues must be taken into account. Other important point is that not every desired final correlation structure is mapped onto a feasible intermediate correlation structure.

3.1 Correlated extremely heterogeneous clutter

For the case of the $\mathcal{G}$ distribution the inverse transform method gives accurate results, Bustos et al. (2009), and it is the method implemented in the toolbox. We directly generate data that describes the return amplitude image, as an example.

Definition 1 *We say that Z_A, the return amplitude image, is a $\mathcal{G}_A^0(\alpha, \gamma, n)$ stochastic process with correlation function ρ_{Z_A} (in symbols $Z_A \sim (\mathcal{G}_A^0(\alpha, \gamma, n), \rho_{Z_A})$) if for all $0 \leq i, j, k, \ell \leq N - 1$ holds that*

1. *$Z_A(k, \ell)$ obeys a $\mathcal{G}_A^0(\alpha, \gamma, n)$ law;*

2. *the mean field is $\mu_{Z_A} = E(Z_A(k, \ell))$;*

3. *the variance field is $\sigma_{Z_A}^2 = Var(Z_A(k, \ell))$;*

4. *the correlation function is*
   $$\rho_{Z_A}((i, j), (k, \ell)) = \left(E(Z_A(i, j) Z_A(k, \ell)) - \mu_{Z_A}^2 \right) / \sigma_{Z_A}^2.$$

The scale property of the parameter γ implies that correlation function ρ_{Z_A} and γ are unrelated and, therefore, it is enough to generate a $Z_A^1 \sim (\mathcal{G}_A^0(\alpha, 1, n), \rho_{Z_A})$ field and then simply multiply every outcome by $\gamma^{1/2}$ to get the desired field.

The transformation method for this case consists of the following steps:

1. propose a correlation structure for the $\mathcal{G}_A^0$ field, say, the function ρ_{Z_A};

2. generate a field of independent identically distributed standard Gaussian observations;

3. compute τ, the correlation structure to be imposed to the Gaussian field from ρ_{Z_A}, and impair it using the Fourier transform without altering the marginal properties;

4. transform the correlated Gaussian field into a field of observations of identically distributed $\mathcal{U}(0, 1)$ random variables, using the cumulative distribution function of the Gaussian distribution (Φ);

5. transform the uniform observations into $\mathcal{G}_A^0$ outcomes, using the inverse of the cumulative distribution function of the $\mathcal{G}_A^0$ distribution (G^{-1}).

The function that relates ρ_{Z_A} and τ is computed using numerical tools. In principle, there are no restrictions on the possible roughness parameters values that can be obtained by this method, but issues related to machine precision must be taken into account.

Examples shown in Bustos et al. (2009) were implemented in IDL 5.2 with auxiliary Fortran routines. Our toolbox written in R reproduce their results and generate other set of correlation functions.

3.2 Correlated heterogeneous clutter

Mejail et al. (2003) have shown that the $\mathcal{G}_A^0$ amplitude distribution constitutes a modeling improvement with respect to the widespread $\mathcal{K}_A$ distribution when fitting urban, forested, and deforested areas in remote sensing data. Nevertheless, in the case of correlated deviates, restrictions are imposed by the transformation method. An important issue is that not every desired final correlation structure ρ_{Z_A} is mapped onto a feasible intermediate correlation structure τ.

Thus, for the simulation of heterogeneous texture, a faster and more accurate simulation method was introduced in Flesia (1999) and discussed in Bustos et al. (2001). It involves the use of convolution filters for the generation of gamma deviates, using independent normal random variables as input. This is the method that is implemented in our toolbox for several correlation structures. The procedure for generated heterogeneous return data can be outlined as

1. Generate independent normal observations.

2. Choose the correlation as the square of a suitable function E, defined on $\mathbf{Z}^2$.

3. Calculate the mask θ that the convolution filter will use, such that $\theta * \theta = E$.

4. Apply the convolution filter to the independent normal deviates, obtaining outcomes from the processes with correlation E in each component.

5. Generate the correlated backscatter σ as the sum of the squares of each normal deviate.

6. Generate independent random variables identically distributed as $\Gamma(n, n)$, where n is the desired equivalent number of looks, Y.

7. Return $Z = \sigma.Y$.

4 Experiments

4.1 Simulating a correlated SAR classification phantom

In practice both parametric and non-parametric correlation structures are of interest. The former rely on analytic forms for ρ, while the latter merely specify values for the correlation. Parametric forms for the correlation structure are simpler to specify, and its inference amounts to estimating a few numerical values; non-parametric forms do not suffer from lack of adequacy, but demand the specification (and possibly the estimation) of potentially large sets of parameters.

In the following examples the techniques presented above will be used to generate samples from parametric correlation structures.

4.2 Example 1

We simulated regions of $\mathcal{K}_I$ distributed clutter with correlation structure given by three different characteristic functions,
Gaussian:

$$E(s) = \exp(\frac{-s^2}{2\ell^2}) \tag{13}$$

Exponential:

$$E(s) = \exp(\frac{|s|}{\ell}) \tag{14}$$

Sync

$$E(s) = \exp(\frac{\sin(\ell s/2)}{\ell s/2}) \tag{15}$$

In Figure 1 we show a phantom with six classes, and simulations of *Gamma* and $\mathcal{K}$ clutter with and without correlation, with the parameters given in Tables 2 and 3.

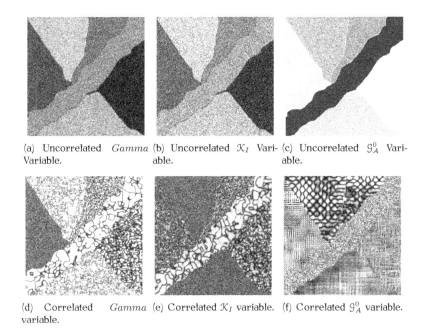

(a) Uncorrelated *Gamma* Variable.

(b) Uncorrelated $\mathcal{K}_I$ Variable.

(c) Uncorrelated $\mathcal{G}_A^0$ Variable.

(d) Correlated *Gamma* variable.

(e) Correlated $\mathcal{K}_I$ variable.

(f) Correlated $\mathcal{G}_A^0$ variable.

Figure 1: Simulated data base example. Panels (a) and (d) uncorrelated and correlated *Gamma* field. Panels (b) and (e) uncorrelated and correlated $\mathcal{K}_I$ field. Panels (c) and (f),uncorrelated and correlated $\mathcal{G}_A^0$ field. Parameters are given in Tables 2 and 3.

	Class 1	Class 2	Class 3	Class 4	Class 5	Class 6
α	0.5	1	1.5	2	2.5	3
ℓ	4	4	2	8	8	8
correlation	Gaussian	Sinc	Gaussian	Gaussian	Exponencial	Exponencial

Table 2: Parameters of clutter phantom depicted in Figure 1, panels (a) for *Gamma* and (b) for $\mathcal{K}_I$ distributed clutter, panels (d) for correlated *Gamma* and (e) for correlated $\mathcal{K}_I$.

4.3   Example 2

We simulated regions of $\mathcal{G}$ distributed clutter with correlation structure given by the following model, which is very popular in applications. Consider $L \geq 2$ an even integer, $0 < a < 1$, $0 < \varepsilon$ (for example $\varepsilon = 0.001$),

	Class 1	Class 2	Class 3	Class 4	Class 5	Class 6
α	-2.5	-3	-3	-9	-9	-1,5
ℓ	3	10	1	10	12	3
correlation	Case 4	Case 2	Case 1	Case 8	Case 6	Case 4

Table 3: Parameters of $\mathcal{G}_A^0$ distributed clutter depicted in Figure 1, panel (c), and correlated clutter depicted in panel (f).

$\alpha < -1$ and $n \geq 1$. Let $h \colon \mathbb{R} \longrightarrow \mathbb{R}$ be defined by

$$h(x) = \begin{cases} x & \text{if } |x| \geq \varepsilon, \\ 0 & \text{if } |x| < \varepsilon. \end{cases}$$

Let ρ_k be defined by

$$\rho_1(k,j) = \begin{cases} h(a\exp(-k^2/\ell^2)) & \text{if } k \geq j, \\ -h(a\exp(-j^2/\ell^2)) & \text{if } k < j. \end{cases} \quad \rho_2(k,j) = \begin{cases} h(a\cos(-k^2/2\ell^2)) & \text{if } k \geq j, \\ -h(a\cos(-j^2/2\ell^2)) & \text{if } k < j. \end{cases}$$
(16)

$$\rho_3(k,j) = \begin{cases} h(a\sin(\pi j/\ell^2)) & \text{if } k \geq j, \\ -h(a\sin(\pi k/\ell^2)) & \text{if } k < j. \end{cases} \quad \rho_4(k,j) = \begin{cases} h(a\sin(4\pi j/\ell^2)) & \text{if } k \geq j, \\ -h(a\sin(4\pi k/\ell^2)) & \text{if } k < j. \end{cases}$$
(17)

$$\rho_5(k,j) = \begin{cases} h(\sin(j\ell)) & \text{if } k \geq j, \\ -h(\sin(k\ell)) & \text{if } k < j. \end{cases}$$
(18)

and $\rho_k(0,0) = 1$.

In Figure 1, two images of size 480×480 each obtained assuming $\gamma = 1.0$, $n = 3$ and different values of a, ℓ, α and correlation functions

Correlation function are different for each class, allowing to insert texture in the classes. Non parametric classification of textured images has a higher accuracy than non-correlated ones, since the classes have more differences. This simple example shows that better classification schemes can be devised if correlation is taken into account.

4.4 Synthetic data classification analysis

Unsupervised classification (also known as clustering) is a method of partitioning image data to extract land-cover information. Unsupervised classification require less input information from the analyst compared to supervised classification because clustering does not require training data. From this process, a map with k classes is obtained. There are hundreds of clustering algorithms. Two of the most conceptually simple family of algorithms are the model-based clustering and distance-based clustering. The model-based approach consists in using certain models for clusters and attempting to optimize the fit between the data and the model. In practice, each cluster can be mathematically represented by a parametric distribution,and the entire data set is therefore modeled by a mixture of these distributions. The most widely used clustering method of this kind is the one based on learning a mixture of Gaussians, with their parameters estimated automatically with the Expectation-Maximization algorithm. The distance based clustering algorithms attempt to minimize an objective function over the set of possible cluster configurations. Of this kind, the two most cited algorithms in the remote sensing literature are k-means and ISODATA clustering algorithms, see Jensen (2005) and Mather (2004) for details.

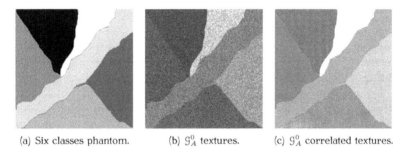

(a) Six classes phantom.   (b) $\mathcal{G}_A^0$ textures.   (c) $\mathcal{G}_A^0$ correlated textures.

Figure 2: Simulated data base example with parameters given in Table 4.

	Class 1	Class 2	Class 3	Class 4	Class 5	Class 6
α	-1.5	-3	-5	-9	-11	-15
ℓ	4	8	8	4	4	4
correlation	Case 4	Case 2	Case 2	Case 8	Case 6	Case 4

Table 4: Correlated $\mathcal{G}_A^0$ distributions considered in the phantom of Figure 2.

All these procedures consider spectral information as independent draws of the underlying joint density. Spatial correlation is often reinforced by the use of a hidden Markovian model on the labeling field. ICM (iterated conditional modes) is a procedure that estimates the best clustering map that fits the hidden Markovian model, usually a eight neighbors Potts model, see Frery et al. (2009), Gimenez et al. (2014) for details on the method. Initial points can be given by the k-means algorithm or the EM-MG (mixture of Gaussians) algorithm. Mejail et al. (2003) made an important Monte Carlo experiment with simulated $\mathcal{G}_A^0$ data, classifying the simulated return imagery with ICM using as starting point a clustering map made over pointwise estimations of the roughness parameter α of the $\mathcal{G}_A^0$. Parameter estimation is usually a sore spot when considering $\mathcal{G}$ distributions, Lucini (2002) discussed the numerical problems of maximum likelihood and robust estimation methods, and the poor accuracy of the moment method. Since we want only to stress the importance of considering correlation properties in simulation studies involving $\mathcal{G}_A^0$ distributions, we report differences on clustering accuracy on nonparametric methods, computed over simulated image returns with the same parameters and different correlation properties.

In this section we show a small simulation example involving automatic pointwise clustering algorithms, k-means, ISODATA, EM-MG as starting point of contextual ICM in the case of uncorrelated $\mathcal{G}_A^0$ data, and without contextual ICM in the case of correlated $\mathcal{G}_A^0$ data.

Frery et al. (2009) reports a Monte Carlo experiment to assess the performance of the pointwise and contextual classification procedures with training stage. We follow their design; each replication consists of assuming a certain image class and transforming classes into observations following the assumed $\mathcal{G}_A^0$ and correlated $\mathcal{G}_A^0$ models, producing clustering maps and validating them. We show only results for k-means algorithms, giving the number of clusters $k = 6$ as prior information: k-means as a prior to ICM contextual estimation, as a way of incorporate spacial correlation estimation in the clustering approach, see Frery et al. (2009), and k-means over the correlated model. Examples of the images considered are shown in Figure 2. The parameters used in the simulation are given in Table 3. The accuracy results of ISODATA and EM-MG are similar.

In Figure 3 we observe two clustering results over uncorrelated data, with *kappa* values significatively different (non overlapping 95% confidence intervals). Clustering improves when ICM is applied, reinforcing the idea that spatial or contextual correlation must be considered in the model for a better clustering accuracy. The third image is a clustering map made with k-means over the correlated clutter, simulated with the same parameter than the uncorrelated image, and different correlation properties per class, see Table 3 for details.

The confusion matrix shown in Table 5 shows the change in false positives and false negatives when considering correlation. The correlation lag was chosen large for almost all classes, which helped differentiate the textures from the independent $\mathcal{G}_A^0$ data. Such textures were simulated with

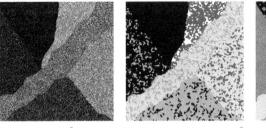

(a) k-means on $\mathcal{G}_A^0$, $kappa =$ (b) k-means-ICM on $\mathcal{G}_A^0$, (c) k-means-ICM on corr.
0.31 $kappa = 0.48$ $\mathcal{G}_A^0$, $kappa = 0.86$

Figure 3: Clustering maps over images in Figure 2.

correlation structure similar to real data as shown in Bustos et al. (2009).

k-means over uncorrelated data					
	A-2	B-2	C-2	D-2	E-2	F-2
A-2	0.7368	0.2465	0.0167	0	0	0
B-2	0.3428	0.4716	0.1775	0.0064	0.0016	0
C-2	0.0860	0.3060	0.3666	0.1365	0.0782	0.0267
D-2	1.0000	0	0	0	0	0
E-2	0.9622	0.0378	0	0	0	0
F-2	0.0109	0.0673	0.1665	0.2315	0.2468	0.2770
k-means over uncorrelated data					
	A-2	B-2	C-2	D-2	E-2	F-2
A-2	1.0000	0	0	0	0	0
B-2	0	0.9878	0.0008	0.0115	0	0
C-2	0	0	0.8655	0.1345	0	0
D-2	0	0	0	0.9200	0.0800	0
E-2	0	0	0	0.2520	0.6957	0.0522
F-2	0	0	0	0	0.0179	0.9821

Table 5: Confusion matrix of methods k-means over correlated and uncorrelated data.

In order to compare how similar two classifications are, it is convenient to summarize the data of the two images in a table. In Table 6 we show similarities between clustering of correlated data using ISODATA and k-means. An overall measure of how similar ISODATA clustering is to k-means clustering can be derived by identifying for each class in classification 1 the class with maximum number of pixels in classification 2. The next step is to calculate the overall percentage of these pixels. For the table below this would equate to:

$$similarity = \sum \max[n_{ij}]\frac{100}{N} \qquad (19)$$

	A-2	B-2	C-2	D-2	E-2	F-2	Max	%Err
A-1	36224.00	947.00	0.00	159.00	0.00	63.00	36224.00	3.13
B-1	3234.00	23962.00	0.00	1.00	0.00	10371.00	23962.00	36.22
C-1	0.00	0.00	20327.00	0.00	111.00	0.00	20327.00	0.54
D-1	14.00	0.00	0.00	51772.00	211.00	0.00	51772.00	0.43
E-1	0.00	0.00	0.00	17.00	43284.00	0.00	43284.00	0.04
F-1	21688.00	95.00	0.00	17280.00	638.00	2.00	21688.00	45.37
Max	36224.00	23962.00	20327.00	51772.00	43284.00	10371.00		
%Err	40.77	4.17	0.00	25.22	2.17	0.62		

Table 6: ISODATA clustering is 85.62 % similar to k-means clustering. k-means clustering is 80.7 % similar to ISODATA clustering. Overall agreement 83.1591 %. Accuracy 0.92. Kappa 0.71

5 Conclusion

In this work we revised several methods for the simulation of correlated clutter with desirable marginal law and correlation structure. These algorithms allow the obtainment of precise and controlled first and second order statistics, and they have been implemented using standard numerical tools in the free software R. The adequacy of the algorithms for the simulation of several scenarios has been assessed within a clustering simulation study involving the use of correlation. Kappa coefficient and confusion matrix have been used as a objective evaluation criteria. The results show that contextual modeling improves significatively the accuracy of the classifier, including correlation as a prior hypothesis on the labeling field or including correlation in the reflectivity data. To continue this work, we are planning a more ambitious Monte Carlo simulation involving classification based on the roughness parameter using correlated data. In our small example introduced here, correlation help separate classes with close mean value, since the distributions are intrinsically different.

Acknowledgment

This work has been partially supported by grants from Secyt-UNC and CONICET. The R code cited in this article can be downloaded from http://www.famaf.unc.edu.ar/~flesia. The code runs in the latest R environment R3.1.2 under Ubuntu12.04. It compiles and runs on 32 and 64 bit platforms. For windows users, isodata.R needs to be changed to support .dll files instead of .so files.

References

Almiron, M. G., Lopes, B., Oliveira, A. L., Medeiros, A. C. & Frery, A. C. (2010), 'On the numerical accuracy of spreadsheets', *Journal of Statistical Software* **34**(4), 1–29.

Bustos, O. H. & Frery, A. C. (2006), 'Statistical functions and procedures in idl 5.6 and 6.0', *Computational statistics & data analysis* **50**(2), 301–310.

Bustos, O. H., Flesia, A. G. & Frery, A. C. (2001), 'Generalized method for sampling spatially correlated heterogeneous speckled imagery', *EURASIP Journal on Applied Signal Processing* **2001**(2), 89–99.

Bustos, O. H., Flesia, A. G., Frery, A. C. & Lucini, M. M. (2009), 'Simulation of spatially correlated clutter fields', *Communications in Statistics - Simulation and Computation* **38**(10), 2134– 2151.

Flesia, A. G. (1999), Caracterización espectral del modelo estocástico para imágenes : estudio y estimación de la densidad espectral de potencia en imágenes SAR, PhD thesis, Facultad de Matemática, Astronomía y Física, Universidad nacional de Córdoba, Argentina.

Frery, A. C., Ferrero, S. & Bustos, O. H. (2009), 'The influence of training errors, context and number of bands in the accuracy of image classification', *International Journal of Remote Sensing* **30**(6), 1425–1440.

Frery, A. C., Muller, H.-J., Yanasse, C. d. C. F. & Sant'Anna, S. J. S. (1997), 'A model for extremely heterogeneous clutter', *Geoscience and Remote Sensing, IEEE Transactions on* **35**(3), 648–659.

Gimenez, J., Frery, A. & Flesia, A. (2014), 'When data do not bring information: A case study in Markov random fields estimation', *Selected Topics in Applied Earth Observations and Remote Sensing, IEEE Journal of,* in press.

Jensen, J. R. (2005), *Introductory Digital Image Processing: A Remote Sensing Perspective*, Upper Saddle River : Pearson Prentice Hall.

Lucini, M. M. (2002), M-estimadores en imágenes de radar de apertura sintética, PhD thesis, Facultad de Matemática, Astronomía y Física, Universidad nacional de Córdoba, Argentina.

Mather, P. (2004), *Computer Processing of Remotely-Sensed Images*, John Wiley & Sons, Ltd.

Mejail, M. E., Frery, A. C., Jacobo-Berlles, J. & Bustos, O. (2001), 'Approximation of distributions for sar images: proposal, evaluation and practical consequences', *Latin American Applied Research* **31**(2), 83–92.

Mejail, M. E., Jacobo-Berlles, J. C., Frery, A. C. & Bustos, O. H. (2003), 'Classification of sar images using a general and tractable multiplicative model', *International Journal of Remote Sensing* **24**(18), 3565–3582.

Ross, S. (2012), *Simulation, fifth edition*, Academic Press.

Uncertainty measures and the concentration of probability density functions[‡]

Hélio Lopes*   Simone Barbosa*

* Departamento de Informática
Pontifícia Universidade Católica do Rio de Janeiro
{lopes,simone}@inf.puc-rio.br

1 Introduction

Uncertainty is a very complex and multifaceted concept. However, several authors agree to see uncertainty as metadata representing lack of knowledge about a model (Viard et al., 2011). Such metadata includes all sort of unknowns, such as: errors, deviations, missing information, or confidence levels, just to cite a few (Potter et al., 2013).

Uncertainty Analysis (UA) aims to quantify the uncertainties in the relevant variables of a model. Nowadays, UA is an indispensable tool to support decisions. Caers (2011) affirms that UA and Decision-Making (DM) should not be considered a sequential set of steps because they, in fact, compose an integral and synergetic process.

Visualization is another important tool that can provide valuable assistance for UA and DM tasks. According to Ware (2013), visualization provides an ability to comprehend huge amounts of data; it allows the perception of emergent properties that were not anticipated; it enables problems with the data to become immediately apparent; and it facilitates not only the understanding of features of the data in different scales, but also the formation of hypotheses.

Uncertainty Visualization (UV) studies how to encode uncertainty information together with the primary data into different graphics primitives (e.g., color, glyph, and texture) in such a way as not to overload the visual perception. In UV applications, data are at least bivariate, because it minimally consists of a primary attribute and its associated degree of

[‡]The authors would like to thank CNPq, PETROBRAS, MICROSOFT, and FINEP for supporting this research. In particular, they thank to Regis Romeu Kruel and Luciano Reis from PETROBRAS for sharing their interest about this topic with us.

uncertainty. Moreover, UV usually deals with multivariate data, because in the majority of cases either primary data or their associated uncertainty are not represented as scalar values (Viard et al., 2011). This provokes the main challenge of UV: to overcome the problem of visualizing the primary data in conjunction with the corresponding uncertainty using a limited number of visual channels, such as: space position, color, texture, opacity, among others. Potter et al. (2012) says that, when we move from quantified uncertainty to visualized uncertainty, we often need to simplify the uncertainty to make it fit into the available visual representation. These challenges have promoted UV to one of the top research problems in visualization for more than one decade now (Mihai & Westermann, 2014; Johnson & Sanderson, 2003).

Contributions. This work aims to describe some simple measures that can be used in visualization applications to quantify uncertainty on discrete and continuous random variables.

Paper outline. Section 2 proposes the use of concentration measures of probability density functions to quantify uncertainty. Section 3 describes two measures of uncertainty for discrete random variables, one based on entropy and other based on statistical distances. Section 4 discusses other concentration measures for continuous random variables. Finally, Section 5 makes some final remarks.

2 Uncertainty as a concentration measure

Let us consider an experiment with an urn filled with colored balls. Suppose that we know the number of colors and their proportion on the urn. Consider that we draw a ball, after having shaken the urn. There are two extreme situations for predicting the color of the drawn ball: when there exists only one color or when the various colors have the same proportion. In the first case, we are completely certain about the outcome, and in the second case, we are completely uncertain about the outcome. Based on these situations, one can suggest a creation of a "**degree of predictability**", which varies between these two extreme cases. Moreover, one can look to the opposite direction of this degree's scale, and then create another measure that represents an "**amount of uncertainty**".

Following the work of Uffink (1990), we now give a precise definition of this idea of amount of uncertainty, because an urn filled with colored balls was used as an illustration in a general experiment.

Suppose that we are dealing with an arrangement in which trials, with a fixed set of possible results, can be performed. On each trial, one and only one of these possible results is observed. Moreover, we assume that certain information about the arrangement is given. Such information is represented by the probability density function over the set of possible results, and it is assumed that it gives a complete description of the arrangement, i.e., no other information is available or relevant to the observation.

In order to associate a degree of predictability or an amount of uncertainty to an experiment, we can use the probability density function (PDF) over the possible results of the arrangement. If the PDF is sharply concentrated, then the outcome of the experiment is highly predictable, or lowly uncertain. On the other way round, if the PDF is uniformly distributed (or non-concentrated) over all possible results, then the outcome of the experiment is lowly predictable, or highly uncertain. The next two sections will describe some concentration measures of PDFs that can be used to measure the amount of uncertainty.

3 Discrete random variables

3.1 Entropies

In information theory, *entropy* measures the unpredictability of information content (Cover & Thomas, 2012). In fact, it quantifies the uncertainty associated with a discrete random variable X with N possible outcomes in $\Omega_N = \{x_1, \ldots, x_N\}$ and whose probability distribution is $P = \{p_1, \ldots, p_N\}$, where p_i indicates $Pr[X = x_i]$.

The most commonly used entropy measure is Shannon entropy (Shannon & Weaver, 1963):

$$H^{(S)}[P] = -\sum_{j=1}^{N} p_j \ln(p_j).$$

Notice that $H^{(S)}$ has a maximum value when the distribution P is uniform. The uniform distribution will from now on be denoted by P_e. For quantification purposes, it is suitable to have a measure of uncertainty in the unit interval $[0, 1]$, so we could use the normalized version of the Shannon entropy measure as follows:

$$\mathcal{H}^{(S)}[P] = H^{(S)}[P]/H^{(S)}[P_e].$$

There are other normalized entropy functionals that could be used as uncertainty measures. Tsallis (Tsallis, 1988) and Rényi (Rényi, 1959) entropies generalize the Shannon entropy according to the following formulas and they are parametrized by the *entropic-index* $q \in \mathbb{R}^+$.

Tsallis: $\mathcal{H}_q^{(T)}[P] = H_q^{(T)}[P]/H^{(T)}[P_e]$, with

$$H_q^{(T)}[P] = \frac{1}{(q-1)} \sum_{i=1}^{N} [p_j - (p_j)^q];$$

Rényi: $\mathcal{H}_q^{(R)}[P] = H_q^{(R)}[P]/H^{(R)}[P_e]$, with

$$H_q^{(R)}[P] = \frac{1}{(1-q)} \ln \left(\sum_{i=1}^{N} (p_j)^q \right).$$

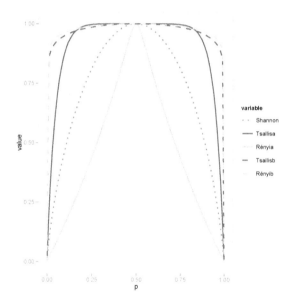

Figure 1: Shannon $\mathcal{H}^{(S)}[P]$, Tsallis $\mathcal{H}_{20}^{(T)}[P]$ (Tsallisa curve) and Tsallis $\mathcal{H}_{0.05}^{(T)}[P]$ (Tsallisb curve), Rényi $\mathcal{H}_{20}^{(R)}[P]$ (Rényia curve) and Rényi $\mathcal{H}_{0.05}^{(R)}[P]$ (Rényib curve) entropies for $P = \{1-p, p\}$ and varying p from 0 to 1.

These two functionals reduce to the corresponding Shannon entropy as q approaches 1. When q has a large positive value these entropies are more sensitive to events that happen frequently. In contrast, for very small values of q, these measures are more sensitive to events that occur rarely.

Figure 1 illustrates the shape of the normalized Shannon $\mathcal{H}^{(S)}[P]$, Tsallis $\mathcal{H}_q^{(T)}[P]$ and Rényi $\mathcal{H}_q^{(R)}[P]$ entropies considering P as the distribution of a random variable obeying a $Bernoulli(p)$ law. In this figure, we vary p from 0 to 1 and select two entropic-indexes for the Tsallis and Rényi entropies: $q = 20$ (Tsallisa and Rényia curves) and $q = 0.05$ (Tsallisb and Rényib curves).

Table 1 illustrates, as a colored map, the normalized Shannon $\mathcal{H}^{(S)}[P]$, Tsallis $\mathcal{H}_q^{(T)}[P]$ and Rényi $\mathcal{H}_q^{(R)}[P]$ entropies considering the distribution $P = \{p_1, p_2, 1 - p_1 - p_2\}$. In this figure, we vary p_1 and p_2 from 0 to 1 and select two entropic-indexes for the Tsallis and Rényi entropies: $q = 20$ and $q = 0.05$.

Maszczyk and Duch (Maszczyk & Duch, 2008) presented an interesting application of these entropies to the construction of decision trees. According to them, Shannon entropy does not distinguish a weak signal overlapping with much a stronger one. Moreover, they say that entropy measures determined by powers of probability, such as Tsallis and Renyi, provide such control.

3.2 Statistical distances

In several information theory applications it is necessary to measure the difference between two probability distributions P and Q (Liese & Vajda, 2006). If we choose Q to be the uniform distribution P_e, the "equilibrium" distribution, we could use the difference between P and P_e to quantify the "disequilibrium" between the possible states of the random variable X whose density is P. The difference between P and P_e, denoted by $D(P||P_e)$, being different from zero reflects that there exist "privileged" (i.e., "more likely") states among the possible ones. Thus, this "disequilibrium" value can be also used as an uncertainty measure.

There are several formulations that can be used to measure the difference between the distributions $P = \{p_1, \ldots, p_N\}$ and $Q = \{q_1, \ldots, q_N\}$, among them we would like to discuss the Euclidian, the Wooters distance (Wootters, 1981), the Kullback-Leibler Shannon entropies (Kullback & Leibler, 1951) and the Jensen-Shannon divergence (Lin, 1991).

A very simple way to compute the difference between P and Q is to use the Euclidean distance $D^{(E)}(P||Q)$:

$$D^{(E)}(P||Q) = \sum_{i=1}^{N} (p_i - q_i)^2.$$

According to Wootters (1981) the use of this straightforward formula ignores the fact that we are dealing with a space of probability distributions. To solve this, he proposed another difference measure (Wootters, 1981), $D^{(W)}(P||Q)$, called the Wooters statistical distance, calculated as follows:

$$D^{(W)}(P||Q) = \cos^{-1}\left(\sum_{i=1}^{N}(p_i)^{1/2}(q_i)^{1/2}\right).$$

Following Kowalski et al. (2011), there are two other difference measures between two distributions that are commonly used in the literature, which are derived from the entropy functionals. The first one uses the relative entropies of the given distributions and is named the Kullback-Leibler-Shannon relative entropy (Kullback & Leibler, 1951):

$$D^{(K)}(P||Q) = \sum_{i=1}^{N} p_i \ln\left(\frac{p_i}{q_i}\right).$$

Notice that the Kullback-Leibler-Shannon measure is not symmetric. The second difference measure uses the entropic difference between P and Q, and is called the Jensen-Shannon divergence (Lin, 1991):

$$D^{(J)}(P||Q) = H^{(S)}\left[\frac{P+Q}{2}\right] - \frac{H^{(S)}[P]}{2} - \frac{H^{(S)}[Q]}{2}.$$

This divergence has an interesting property: its square root is a metric, i.e., it obeys the axioms of distance in a metric space.

As mentioned in the beginning of this section, all these statistical distance measures can be used to quantify uncertainty in a similar way as the entropies. A simple way to do that is to use the following formula:

$$\mathcal{Q}^{(d)} = 1 - \frac{D^{(d)}(P||P_e)}{D^{(d)}(P_{max}||P_e)},$$

with $d \in \{E, W, K, J\}$ representing the Euclidean distance, Wooters statistical distance, Kullback-Leibler-Shannon relative entropy and Jensen-Shannon divergence, respectively; and P_{max} being the distribution that has one entry p_i equal to 1 and the remaining entries all equal to 0. Figure 2 illustrates the shapes of the $\mathcal{Q}^{(\cdot)}$ uncertainty measures by considering P as a Bernoulli(p) distribution, i.e., $P = \{1 - p, p\}$ and varying p from 0 to 1.

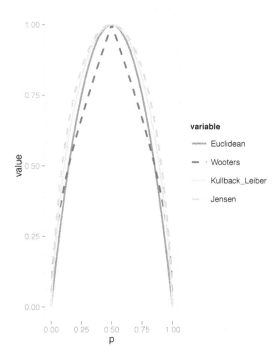

Figure 2: Uncertainty measure $\mathcal{Q}^{(\cdot)}$ using the Euclidean distance, Wooters statistical distance, Kullback-Leibler-Shannon relative entropy and Jensen-Shannon divergence from $P = \{1 - p, p\}$ to $P_e = \{1/2, 1/2\}$ and varying p from 0 to 1.

Table 2 illustrates, as a colored map, the Euclidean distance, Wooters statistical distance, Kullback-Leibler-Shannon relative entropy and Jensen-Shannon divergence from $P = \{p_1, p_2, 1 - p_1 - p_2\}$ to $P_e = \{1/3, 1/3, 1/3\}$ varying p_1 and p_2 from 0 to 1.

4   Continuous random variables

For continuous random variables there are some commonly used con-centration measures of a PDF $p(x)$ of a random variable X. Intuitively, sharply concentrated continuous PDFs are more predictable than PDFs that are spread out over a large interval. An extreme example of a sharply concentrated PDF is when $p(x)$ is the Delta of Dirac function. In this section, we list a few uncertainty measures, as suggested by Uffink (1990). The higher the value of these measures, the more uncertain the outcome of X.

4.1 Standard deviation

The first concentration measure to be mentioned is the *standard deviation* σ, given by:

$$\sigma = \sqrt{E[X^2] - (E[X])^2},$$

where

$$E[f(X)] = \int_{-\infty}^{\infty} f(x) \; p(x) \; dx.$$

4.2 Interquartile range

A second measure of statistical dispersion that could be used is the *Interquartile Range* (IQR) (Zwillinger & Kokoska, 2010). It is equal to the difference between the 3rd and 1st quartiles, i.e., $IQR = Q3 - Q1$, where $Q3 = x : \int_{-\infty}^{x} p(t) \; dt = 0.75$ and $Q1 = x : \int_{-\infty}^{x} p(t) \; dt = 0.25$. In statistics, IQR is considered a robust measure of scale, which means that it is a robust statistic that quantifies statistical dispersion. The IQR is often used to detect outliers in data. An outlier is an observation that falls below $Q1 - 1.5(IQR)$ or above $Q3 + 1.5(IQR)$.

4.3 Half-width

The *Half-Width* (HW) is defined as the distance between the point x_M where the density curve p attains its maximum value M and the nearest point x on the domain of p such that the density curve reaches half of M.

4.4 Equivalence width

The *Equivalence Width* (EW) is defined as the width of the rectangle that has the same area and height as the density curve p:

$$EW = \frac{1}{\max_x\{p(x)\}}.$$

4.5 Shannon entropy

The *Shannon entropy* continuous analogue is also commonly used as a measure of uncertainty associated with a probability density function p of a continuous random variable. It is given by the following formula:

$$H^{(S)}[p] = - \int_{-\infty}^{\infty} p(x) \ln(p(x)) \ dx.$$

4.6 Remarks

In the continuous case, there are some observations that have to be mentioned before choosing a particular measure, such as the ones mentioned here. First of all, it is important to note that these continuous case expressions are mathematically very different from the ones described for the discrete case. This is because they depend not only on the probability distribution, but also on the circumstance that the possible outcomes are represented as real numbers Uffink (1990).

Considering this list of measures, the standard deviation shows to be the most usual choice, except in the case when the random variable obeys the Cauchy law $\left(p(x) = \frac{\gamma}{\pi} \frac{1}{(x-a)^2+\gamma^2} \right)$, in this case the Half-Width is the common choice. This is because the mean and the variance of a random variable obeying a Cauchy law diverge,[1] although this distribution is sharply peaked when γ is small.

Uffink (1990) noticed that all these expressions are invariant under a translation. Moreover, he also observed that if the density is compressed without changing its form, all the above measures will decrease. This means that if we replace $p(x)$ with $\hat{p}(x) = a \cdot p(a \cdot x)$, with $a > 0$, the concentration measure will be a decreasing function of a. However, the continuous analogue for the Shannon entropy for such transformation has a strange behavior. With this expression:

$$H^{(S)}(\hat{p}) = H^{(S)}(p) - \ln(a),$$

one can observe that $H^{(S)}(\hat{p})$ may assume negative values. As a consequence, this expression sometimes is rejected as an uncertainty measure. Other important fact about the Shannon entropy continuous analogue pointed out by Uffink (1990) is that whenever the values of x, like length or mass, is associated to a physical dimension, the unit of the $H^{(S)}$ measure

---

[1]It is interesting to notice that Cauchy created this distribution mainly because of these facts.

will be log-meters or log-kilograms. To solve these two issues, one could replace $H^{(S)}$ by $\exp\left(H^{(S)}\right)$.

Finally, some special care has to be taken when the density function contains more than one peak or a singularity. In these cases, for example, the half-width is not uniquely defined.

5 Conclusions

In this work we listed a set of possible measures that could be used to quantify uncertainty. These measures somehow quantify the concentration of the PDF curve. We conclude by quoting Cramér (1946), who stated that "all measures of location and dispersion, and of similar properties, are to a large extent arbitrary. This is quite natural, since the properties to be described by such parameters are too vaguely defined to admit of unique measurement by means of a single number. Each measure has advantages and disadvantages of its own, and a measure which renders excellent service in one case may be more or less useless in another." (pp.181-2)

References

Caers, J. (2011), *Modeling Uncertainty in the Earth Sciences*, 1st ed., Wiley-Blackwell.

Cover, T. M. & Thomas, J. A. (2012), *Elements of information theory*, John Wiley & Sons.

Cramér, H. (1946), *Mathematical methods of statistics*, Princeton university press.

Johnson, C. R. & Sanderson, A. R. (2003), 'A next step: Visualizing errors and uncertainty', *IEEE Computer Graphics and Applications* 23(5), 6–10.

Kowalski, A. M., Martín, M. T., Plastino, A., Rosso, O. A. & Casas, M. (2011), 'Distances in probability space and the statistical complexity setup', *Entropy* 13(6), 1055–1075.

Kullback, S. & Leibler, R. (1951), 'On information and sufficiency', *The Annals of Mathematical Statistics* pp. 79–86.

Liese, F. & Vajda, I. (2006), 'On divergences and informations in statistics and information theory', *IEEE Transactions on Information Theory* 52(10), 4394–4412.

Lin, J. (1991), 'Divergence measures based on the shannon entropy', *Information Theory, IEEE Transactions on* **37**(1), 145–151.

Maszczyk, T. & Duch, W. (2008), Comparison of shannon, renyi and tsallis entropy used in decision trees, *in* L. Rutkowski, R. Tadeusiewicz, L. Zadeh & J. Zurada, eds, 'Artificial Intelligence and Soft Computing - ICAISC 2008', Vol. 5097 of *Lecture Notes in Computer Science*, Springer Berlin Heidelberg, pp. 643–651.

Mihai, M. & Westermann, R. (2014), 'Visualizing the stability of critical points in uncertain scalar fields', *Computers & Graphics* **41**(0), 13–25.

Potter, K., Gerber, S. & Anderson, E. W. (2013), 'Visualization of uncertainty without a mean', *IEEE Computer Graphics and Applications* **33**(1), 75–79.

Potter, K., Rosen, P. & Johnson, C. R. (2012), From quantification to visualization: A taxonomy of uncertainty visualization approaches, *in* 'Uncertainty Quantification in Scientific Computing', Springer, pp. 226–249.

Rényi, A. (1959), 'On the dimension and entropy of probability distributions', *Acta Mathematica Academiae Scientiarum Hungarica* **10**(1-2), 193–215.

Shannon, C. E. & Weaver, W. (1963), *Mathematical theory of communication*, University Illinois Press.

Tsallis, C. (1988), 'Possible generalization of boltzmann-gibbs statistics', *Journal of Statistical Physics* **52**(1-2), 479–487. ISSN 0022-4715.

Uffink, J. B. M. (1990), 'Measures of uncertainty and the uncertainty principle'.

Viard, T., Caumon, G. & Lévy, B. (2011), 'Adjacent versus coincident representations of geospatial uncertainty: Which promote better decisions?', *Computers & Geosciences* **37**(4), 511–520.

Ware, C. (2013), *Information visualization: perception for design*, Elsevier.

Wootters, W. K. (1981), 'Statistical distance and hilbert space', *Physical Review D* **23**(2), 357–362.

Zwillinger, D. & Kokoska, S. (2010), *CRC standard probability and statistics tables and formulae*, CRC Press.

Table 1: Shannon $\mathcal{H}^{(S)}[P]$, Tsallis $\mathcal{H}_q^{(T)}[P]$ and Rényi $\mathcal{H}_q^{(R)}[P]$ entropies considering the distribution $P = \{p_1, p_2, 1 - p_1 - p_2\}$.

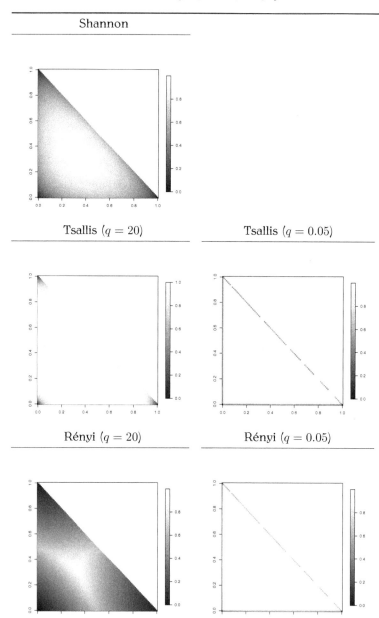

Table 2: the Euclidean distance, Wooters statistical distance, Kullback-Leibler-Shannon relative entropy and Jensen-Shannon divergence from $P = \{p_1, p_2, 1 - p_1 - p_2\}$ to $P_e = \{1/3, 1/3, 1/3\}$ varying the values of p_1 and p_2 from 0 to 1.

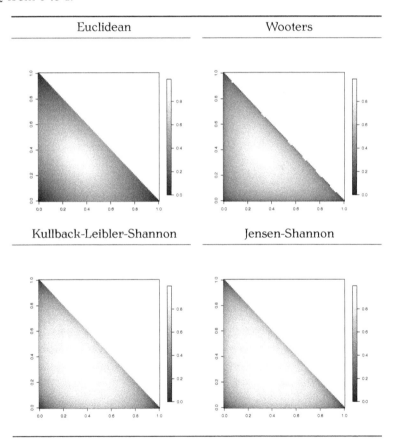

Using stochastic distances on polarimetric SAR image processing

Leonardo Torres*   Sidnei J. S. Sant'Anna*

Corina C. Freitas*   Wagner B. Silva[†]   Alejandro C. Frery[‡]

* Instituto Nacional de Pesquisas Espaciais
{ljmtorres, sidnei, corina}@dpi.inpe.br

[†] Instituto Militar de Engenharia
wbarreto.w3@gmail.com

[‡] Universidade Federal de Alagoas
acfrery@gmail.com

1 Introduction

Synthetic Aperture Radar (SAR) data are generated by a coherent illuminated system. It is known that these data incorporate a granular noise, denominated speckle noise, that degrades its quality, and it is also present in the laser, ultrasound-B, and sonar imagery (Goodman, 1976). Digital image tasks such as segmentation, extraction, analysis and, classification of objects in the image can be extremely influenced by this kind of noise.

Statistical approach is essential for dealing with speckled data. It provides comprehensive support for developing procedures for interpreting the data efficiently, and to simulate plausible images (Gao, 2010). In this paper, the multiplicative model was used to describe the speckle noise (see Section 2). Different statistical distributions are proposed in the literature to describe speckle data (Frery et al., 1997). Among these distributions, the Wishart model is widely accepted to describe homogeneous areas. Based on this distribution, and using the Information Theory framework, Nascimento (2012) has developed several stochastic distances and their corresponding hypothesis tests.

This article presents two applications of the use of stochastic distances in polarimetric SAR image processing. The first application is about speckle noise filtering, and the second deals with the classification of regions of these images. The noise filter presented here, denominated Stochastic Distances Nonlocal Means (SDNLM), is an adaptive nonlinear extension of the NL-means algorithm introduced by Buades et al. (2005). The region classifier, denominated here PolClass, performs a supervised classification, in which a stochastic distance of each image region to training samples (representing the defined classes) is computed. The classification of the regions is made using statistical tests involving these distances. An important feature of this classifier is that, in addition to the classified image, a precision map of the classification is also provided.

2 Polarimetric SAR (PolSAR) image statistical properties

The microwave imagery is usually performed by the transmission and reception of linearly polarized signals. In particular, in full-polarimetric sensors, four combinations of these polarizations are recorded, leading to four backscattering components: S_{HH} (horizontal-horizontal), S_{HV} (horizontal-vertical), S_{VH} (vertical-horizontal), and S_{VV} (vertical-vertical). Under certain assumptions (Ulaby & Elachi, 1990), the reciprocity theorem is satisfied and it can be assumed that $S_{HV} = S_{VH}$. In general, we may consider systems with m polarization elements, which constitute a complex random vector denoted by:

$$y = [S_1 \ S_2 \ \cdots \ S_m]^\top, \tag{1}$$

where $(\cdot)^\top$ is the transposition operator. It is commonly assumed that the scattering vector y follows a circular complex Gaussian law (Goudail & Réfrégier, 2004).

Multilook PolSAR data are usually generated in order to enhance the signal-to-noise ratio (SNR):

$$Z = \frac{1}{L} \sum_{i=1}^{L} y_i y_i^H, \tag{2}$$

where $(\cdot)^H$ is the Hermitian operator, y_i represents the scattering vector in the ith look, and L is the number of looks, a parameter related to the noise effect in SAR imagery. The matrix Z is defined over the set of positive-definite Hermitian matrices $\mathcal{A}$. Moreover, Goodman (1963) showed that LZ follows the ordinary complex Wishart law and, therefore, the density of the scaled random matrix Z is

$$f_Z(Z'; \Sigma, L) = \frac{L^{mL}|Z'|^{L-m}}{|\Sigma|^L \Gamma_m(L)} \exp\left[- L \operatorname{tr}(\Sigma^{-1} Z') \right], \tag{3}$$

where m is the order of $\boldsymbol{\Sigma}$, $|\cdot|$ represents the determinant operator, $\mathrm{tr}(\cdot)$ is the trace of matrix, $m \leq L$, $\Gamma_m(L) = \pi^{m(m-1)/2} \prod_{k=0}^{m-1} \Gamma(L-k)$, $\boldsymbol{\Sigma} = \mathrm{E}\{\boldsymbol{Z}\}$, and $\mathrm{E}\{\cdot\}$ is the expectation operator.

3   Stochastic distances

Assume that $\boldsymbol{X}$ and $\boldsymbol{Y}$ are random variables associated with density functions $f_{\boldsymbol{X}}(\boldsymbol{Z}'; \boldsymbol{\theta}_1)$ and $f_{\boldsymbol{Y}}(\boldsymbol{Z}'; \boldsymbol{\theta}_2)$, respectively, where $\boldsymbol{\theta}_1$ and $\boldsymbol{\theta}_2$ are parameter vectors. The random variables are assumed to share a common support $\mathcal{A}$. The (h, ϕ)-divergence between $f_{\boldsymbol{X}}$ and $f_{\boldsymbol{Y}}$ is defined by (Salicrú et al., 1994):

$$D_\phi^h(\boldsymbol{X}, \boldsymbol{Y}) = h\left(\int_{\mathcal{A}} \phi\left(\frac{f_{\boldsymbol{X}}(\boldsymbol{Z}'; \boldsymbol{\theta}_1)}{f_{\boldsymbol{Y}}(\boldsymbol{Z}'; \boldsymbol{\theta}_2)} \right) f_{\boldsymbol{Y}}(\boldsymbol{Z}'; \boldsymbol{\theta}_2)\mathrm{d}\boldsymbol{Z}' \right), \tag{4}$$

where $h\colon (0, \infty) \to [0, \infty)$ is a strictly increasing function with $h(0) = 0$, $\phi\colon (0, \infty) \to [0, \infty)$ is a convex function, and indeterminate forms are assigned the value zero. The differential element $\mathrm{d}\boldsymbol{Z}'$ is given by (Goodman, 1963):

$$\mathrm{d}\boldsymbol{Z}' = \mathrm{d}Z_{11}\mathrm{d}Z_{22} \cdots \mathrm{d}Z_{mm} \underbrace{\prod_{\substack{i, j = 1}}^{m}}_{i<j} \mathrm{d}\Re\{Z_{ij}\}\mathrm{d}\Im\{Z_{ij}\}, \tag{5}$$

where Z_{ij} is the (i, i)-th entry of Z, and operators $\Re\{\cdot\}$ and $\Im\{\cdot\}$ return real and imaginary parts of their arguments, respectively.

Well-known divergences arise from adequate choices of h and ϕ. Among them, we examined the following: (i) Kullback-Leibler (Seghouane & Amari, 2007), (ii) Rényi, (iii) Bhattacharyya (Kailath, 1967), and (iv) Hellinger (Nascimento et al., 2010). As the triangular inequality is not necessarily satisfied, not every divergence measure is a metric (Burbea & Rao, 1982). Additionally, the symmetry property is not followed by some of these divergence measures. Nevertheless, such tools are mathematically appropriate for comparing the distribution of random variables (Jager & Wellner, 2007). The following expression has been suggested as a possible solution for this issue (Seghouane & Amari, 2007):

$$d_\phi^h(\boldsymbol{X}, \boldsymbol{Y}) = \frac{D_\phi^h(\boldsymbol{X}, \boldsymbol{Y}) + D_\phi^h(\boldsymbol{Y}, \boldsymbol{X})}{2}. \tag{6}$$

Functions $d_\phi^h : \mathcal{A} \times \mathcal{A} \to \mathbb{R}$ are distances over $\mathcal{A}$ since, for all $\boldsymbol{X}, \boldsymbol{Y} \in \mathcal{A}$, the following properties hold:

1. $d_\phi^h(\boldsymbol{X}, \boldsymbol{Y}) \geq 0$ (Non-negativity).

2. $d_\phi^h(\boldsymbol{X}, \boldsymbol{Y}) = d_\phi^h(\boldsymbol{Y}, \boldsymbol{X})$ (Symmetry).

3. $d_\phi^h(\boldsymbol{X}, \boldsymbol{Y}) = 0 \Leftrightarrow \boldsymbol{X} = \boldsymbol{Y}$ (Definiteness).

Table 1 shows the h and ϕ functions which lead to the distances considered in this work.

When considering distances between distributions of the same family, only parameters are relevant. In this case, the parameters $\boldsymbol{\theta}_1$ and $\boldsymbol{\theta}_2$ replace the random variables $\boldsymbol{X}$ and $\boldsymbol{Y}$ as arguments of the discussed distances. This notation is in agreement with that of Salicrú et al. (1994). Nascimento (2012) and Frery et al. (2014) derived closed-form expressions for the d_{KL}, d_{R}^β, d_{B}, and d_{H} distances between two relaxed complex Wishart distributions.

In the following we recall the hypothesis test based on stochastic distances proposed by Salicrú et al. (1994). Let M-point vectors $\widehat{\boldsymbol{\theta}}_1 = (\widehat{\theta}_{11}, \ldots, \widehat{\theta}_{1M})$ and $\widehat{\boldsymbol{\theta}}_2 = (\widehat{\theta}_{21}, \ldots, \widehat{\theta}_{2M})$ be the Maximum Likelihood (ML) estimators of parameters $\boldsymbol{\theta}_1$ and $\boldsymbol{\theta}_2$ based on independent samples of sizes N_1 and N_2, respectively. Under the regularity conditions discussed in (Salicrú et al., 1994) the following lemma holds:

Lemma 3.1 *If* $\dfrac{N_1}{N_1+N_2} \xrightarrow[N_1,N_2 \to \infty]{} \lambda \in (0,1)$ *and* $\boldsymbol{\theta}_1 = \boldsymbol{\theta}_2$, *then*

$$S_\phi^h(\widehat{\boldsymbol{\theta}}_1, \widehat{\boldsymbol{\theta}}_2) = \frac{2N_1 N_2}{N_1 + N_2} \frac{d_\phi^h(\widehat{\boldsymbol{\theta}}_1, \widehat{\boldsymbol{\theta}}_2)}{h'(0)\phi''(1)} \xrightarrow[N_1,N_2 \to \infty]{\mathcal{D}} \chi_M^2, \qquad (7)$$

where "$\xrightarrow{\mathcal{D}}$" denotes convergence in distribution, h' and ϕ'' are the first and second derivative functions, and χ_M^2 represents the chi-square distribution with M degrees of freedom.

Based on Lemma 3.1, statistical hypothesis tests for the null hypothesis $\boldsymbol{\theta}_1 = \boldsymbol{\theta}_2$ can be formulated in the following proposition.

Proposition 3.2 *Let N_1 and N_2 be large and $S_\phi^h(\widehat{\boldsymbol{\theta}}_1, \widehat{\boldsymbol{\theta}}_2) = s$, then the null hypothesis $\boldsymbol{\theta}_1 = \boldsymbol{\theta}_2$ can be rejected at level α if $\Pr(\chi_M^2 > s) \leq \alpha$.*

We denote the statistics based on the Kullback-Leibler, Rényi, Bhattacharyya, and Hellinger distances as S_{KL}, S_{R}^β, S_{B}, and S_{H}, respectively. These measures are used here for the construction of methodologies of a noise speckle filter and of a region classifier for polarimetric SAR images, under the assumption of the Wishart model to describe the data.

Table 1: (h, ϕ)-functions.

Classes h-ϕ	used IT measures	$h(y)$	$\phi(x)$
Distances	Kullback-Leibler	$y/2$	$(x-1)\log x$
	Rényi (order β) $0 < \beta < 1$	$\frac{1}{\beta-1}\log((\beta-1)y+1),\ 0 \leq y < \frac{1}{1-\beta}$	$\frac{x^{1-\beta}+x^\beta-\beta(x-1)-2}{2(\beta-1)}$
	Bhattacharyya	$-\log(-y+1),\ 0 \leq y < 1$	$-\sqrt{x}+\frac{x+1}{2}$
	Hellinger	$y/2,\ 0 \leq y < 2$	$(\sqrt{x}-1)^2$

4 Applications of stochastic distances to PolSAR image processing

4.1 Nonlocal means (NLM) filters

The NLM filter, in its genuine formulation, is appropriated to reduce the additive white Gaussian noise. Buades et al. (2005), under the idea that spatial structures (patches) repeat, define a neighborhood of a pixel i as any set of pixels j in the image such that a window around j looks like a window around i. All pixels in that neighborhood can be used for predicting the value at i. Consequently, it is possible to do a better pixel estimation compared to local information estimation.

The pixels belonging to the defined neighborhood are used to calculated the weighted average filter. The weights were originally obtained based on the Euclidean distance, which is used to measure the similarity between a central and adjacent regions in a search window. The noise-free pixel is calculated as:

$$g(s,t) = \sum_{u,v \in W} f_I(s+u, t+v)\, w(u,v), \tag{8}$$

where $w(u,v)$ are the weights defined on the search window W. The restored image g is the convolution of the input image f_I with the mask $w = w'/\sum w'(u,v)$. The factor $w'(u,v)$ is inversely proportional to the similarity measure:

$$w'(u,v) = \exp\left\{ -\frac{1}{\tau} \sum_{k \in W} |f_I(\rho_u(k)) - f_I(\rho_v(k))|^2 \right\}, \tag{9}$$

where $\tau > 0$ controls the decreasing function, $\rho_u(k)$ and $\rho_v(k)$ are the observations of the k-th pixel centered on the sample in u and v, respectively.

The NLM filter and its extensions depend on computing the weights of a convolution mask as similarity measure functions: the most similar two samples are (according to a metric), the biggest is the weight and the contribution of the central pixel in the filtering process.

Deledalle et al. (2010, 2013) applied this methodology to PolSAR data using the Kullback-Leibler distance between two complex Gaussian distributions. Additionally, Chen et al. (2011) presented a NLM filter for PolSAR images, however, they considered a complex Wishart distribution, and used the likelihood ratio test, as presented by Conradsen et al. (2003), to compute the weights of the neighborhood pixels.

Zhong et al. (2014) derived a new NLM filter applying some changes on the Lee filter (Lee et al., 1999), called Nonlocal Lee (NL-Lee). The NL-Lee filter is constructed combining two processes: (i) "structural similarity", introduced by the NLM method; and (ii) "homogeneity similarity", introduced by the Lee filter. In calculating the weights, Zhong et al. (2014) used the same formulation given by (Chen et al., 2011) to provide the structural similarity between samples. On the other hand, the homogeneity similarity is provided by a span image using the MMSE criteria, as in (Lee et al., 1999).

4.1.1 Stochastic distances nonlocal means (SDNLM) filter

In the filter proposed by Torres et al. (2014), namely Stochastic Distances Nonlocal Means (SDNLM), three different squared and odd sized windows are defined: (i) central patch, (ii) neighborhood patch and (iii) searching windows. The central patch window is the window around the pixel to be filtered; the neighborhood patch window is the window around the possible neighborhood pixels; and the searching window is the region that contains the possible neighborhood pixels, i.e. the central pixels of the neighborhood patches. In the SDNLM filter the central patch and the neighborhood patch windows are of the same size ($q \times q$ pixels), and the searching window is of size $n \times n$ pixels. The central patch, centered in the pixel denominated Z_1, is thus compared with $N = n^2 - q^2$ neighboring patches, whose center pixels are Z_i, $i = \{2, \ldots, N + 1\}$, as illustrated in Figure 1 for $q = 3$ and $n = 7$. The estimate of the noise-free observation at Z_1 is a weighted sum of the observations at $Z_1, \ldots, Z_{n^2}$. These weights are equal to 1 for the q^2 pixels belonging to the central patch window, and the remaining weights are functions of the p-value ($p(1, i)$) of the statistical test of equality of the parameters of the two Wishart distributions, one related to the central patch window and the other related to the neighborhood patch window around Z_i:

$$w'_{\mathrm{NL}}(1, i) = \begin{cases} 1 & \text{if } p(1, i) \geq \eta, \\ \frac{2}{\eta} p(1, i) - 1 & \text{if } \frac{\eta}{2} < p(1, i) < \eta, \\ 0 & \text{otherwise,} \end{cases} \tag{10}$$

where η is the confidence of the test, chosen by the user. This function is illustrated in Figure 2. In this way we employ a soft threshold instead of an accept-reject decision. This allows the use of more evidence than with a binary decision.

4.2 Polarimetric Region Classifier

Silva (2013) developed and evaluated a polarimetric region classifier based on stochastic distances and hypothesis tests, denominated PolClass classi-

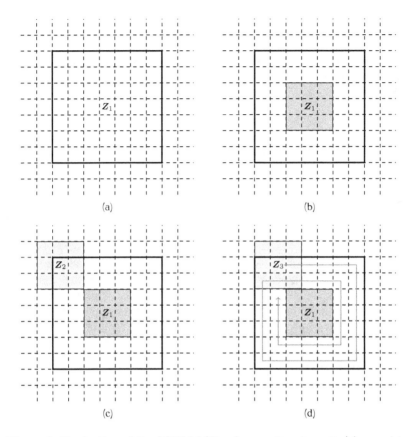

Figure 1: Illustration of the SDNLM filter for $q = 3$ and $n = 7$: (a) searching window; (b) central patch window; (c) first neighborhood patch window; and (d) second neighborhood patch window and the pathway of the remaining patch windows.

fier. This classifier is organized into three modules, divided according to the assumed statistical model and to the image type used. The first module is proposed for PolSAR data classification and assumes the Wishart distribution for covariance matrices modeling. The second one, which is intended for classification of intensity pairs of SAR images, assumes the multilook Intensity-Pair distribution developed by (Lee et al., 1994). The third one, which is intended for classification of multivariate SAR images in amplitude and images from optical sensors, assumes a multivariate

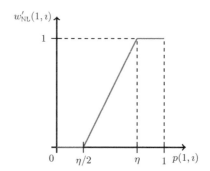

Figure 2: Weight function for every pair of patches $(1, \imath)$, $2 \leq \imath \leq N$.

Gaussian model (Theodoridis & Koutroumbas, 2008). Different stochastic distances are used in each classifier module, such as Bhattacharyya, Kullback-Leibler, Hellinger, Rényi, Chi-Squared, and Triangular distances.

Initially, it is assumed that the PolSAR image is partitioned in r disjoint segments, $C_1, \ldots, C_r$. PolSAR data in each segment are denoted by $\boldsymbol{Z}_{i|C_k}$, where $i = 1, 2, \ldots, N_k$, $k = 1, 2, \ldots, r$ and N_k is the number of pixels in the kth segment. Under the hypothesis that $\boldsymbol{Z}_{1|C_k}, \boldsymbol{Z}_{2|C_k}, \ldots, \boldsymbol{Z}_{N_k|C_k}$ are random sample drawn from $\boldsymbol{Z}_{C_k} \sim \mathcal{W}(\boldsymbol{\Sigma}_k, L)$, for all $k = 1, 2, \ldots, r$, one has that the ML estimator for $\boldsymbol{\Sigma}_k$ is given by $\widehat{\boldsymbol{\Sigma}}_k = N_k^{-1} \sum_{i=1}^{N_k} \boldsymbol{Z}_{i|C_k}$.

Suppose, for a supervised approach, that the user provides P prototypes in the form of samples (training samples) representing the defined classes, and covariances matrices $\widehat{\boldsymbol{\Sigma}}_\ell$, $1 \leq \ell \leq P$, are obtained by ML estimation. The purpose is to classify each segment by its closeness to the training samples. To that end, $r \times P$ test statistics for $\mathcal{H}_0 \colon \boldsymbol{\Sigma}_k = \boldsymbol{\Sigma}_\ell$, $k = 1, 2, \ldots, r$, $\ell = 1, 2, \ldots, P$ are computed using Lemma 3.1 and Proposition 3.2. The classification based on minimum test statistics consists of assigning the segment C_k to the class represented by prototype t if

$$s_\phi^h(\widehat{\boldsymbol{\Sigma}}_k, \widehat{\boldsymbol{\Sigma}}_t) < s_\phi^h(\widehat{\boldsymbol{\Sigma}}_k, \widehat{\boldsymbol{\Sigma}}_\ell) \tag{11}$$

for all $t \neq \ell$, where s_ϕ^h is a sample outcome of S_ϕ^h. Once the segment C_k has been assigned to the class represented by prototype t, the p-value of the statistical test, related to the confidence of the classification of C_k, is computed by:

$$p_{k,t} = \Pr(\chi_v^2 > s_\phi^h(\widehat{\boldsymbol{\Sigma}}_k, \widehat{\boldsymbol{\Sigma}}_t)), \tag{12}$$

where v is the number of parameters of the considered model: $v = m^2$ for the Wishart distribution, $v = 3$ for the Intensity-Pair distribution, and

$v = m(m + 3)/2$ for the m-variate Gaussian distribution. At the end of the classification, two images are provided to the user : the classified image, and the p-values image, which represents the precision map of the classification.

5 Results

5.1 SDNLM filter on PolSAR image

The proposed filter, termed SDNLM (*Stochastic Distances Nonlocal Means*) filter, was compared with three other filters: SMB (Scattering-model-based) proposed by Lee et al. (2006), IDAN (*intensity-driven adaptive-neighborhood*) proposed by Vasile et al. (2006), and Boxcar filters. These filters act on a well defined neighborhood (as our implementation of the SDNLM): a square processing window of size of 5×5 pixels, and the SMB filter is adaptive, as the SDNLM filter. For the SDNLM filter, the searching widow is of size 5×5 pixels and the two patch windows are of size 3×3 pixels.

A National Aeronautics and Space Administration Jet Propulsion Laboratory (NASA/ JPL) Airborne SAR (AIRSAR) image of the San Francisco Bay was used for evaluating the proposed filter, see http://earth.eo.esa.int/polsarpro/datasets.html. The original PolSAR image was generated in the L-band, four nominal looks, and 10×10 m spatial resolution. Figure 3(a) shows the Google Map of the area and the Figure 3(b) shows a 350×350 scene of the image in false color using the Pauli decomposition. It consists of assigning $|S_{HH} + S_{VV}|^2$ to the Red channel, $|S_{HH} - S_{VV}|^2$ to the Green channel, and $2|S_{HV}|^2$ to the Blue channel.

Figure 3(c) shows the effect of the Boxcar filter. Despite the evidence that the despeckling process removes noise, it is also clear that the blurring introduced eliminates useful information as, for instance, the Presidio Golf Course: the curvilinear dark features at the center of the image. Figure 3(d) is the result of applying the SMB filter, which shows a good performance, but some details in the edges are eliminated. In particular, the Mountain Lake, the small brown spot at the center of the image, is blurred, as well as the blocks in the urban area. The results of the IDAN filter and of our proposal with $\eta = 90\%$ are shown in Figures 3(e) and 3(f), respectively. Both filters are able to smooth the image in a selective way, but the SDNLM filter enhances more the signal-to-noise ratio while preserving fine details than the IDAN filter.

Table 2 presents the result of assessing the filters in the three intensity channels by means of the equivalent number of looks (ENL) in the large

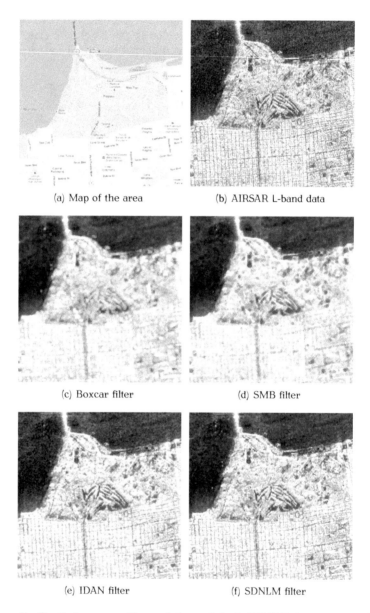

(a) Map of the area   (b) AIRSAR L-band data

(c) Boxcar filter   (d) SMB filter

(e) IDAN filter   (f) SDNLM filter

Figure 3: Pauli decomposition of the original AIRSAR image over San Francisco and its filtered versions.

forest area, a homogeneous area, and of the BRISQUE index proposed by Mittal et al. (2012). The Blind/Referenceless Image Spatial QUality Evaluator (BRISQUE) is a holistic measure of quality on no-reference images. It has a score typically between 0 and 100, where 0 represents the best quality and 100 the worst. This image quality evaluator quantifies possible losses of "naturalness" in the image and does not compute distortions. The results are consistent with those observed by Torres et al. (2014): the mere evaluation of the noise reduction by the ENL suggests the Boxcar filter as the best one, but the natural scene distortion-generic BRISQUE index is better after applying the SDNLM filter.

Table 2: Image quality indexes in the real PolSAR image.

Filter	ENL			BRISQUE Index		
	HH	*HV*	*VV*	*HH*	*HV*	*VV*
Real data	3.867	4.227	4.494	58.258	70.845	61.593
Boxcar	**14.564**	**25.611**	**18.946**	36.498	37.714	36.792
SMB	11.491	20.415	15.407	44.997	51.547	49.412
IDAN	2.994	3.732	3.923	28.823	34.853	34.691
SDNLM 80 %	7.263	11.532	8.299	**27.841**	**27.256**	**33.541**
SDNLM 90 %	8.177	12.404	9.013	28.622	35.622	35.016
SDNLM 99 %	10.828	18.379	13.075	31.026	33.881	36.881

5.2 PolClass on PolSAR image

The classification procedure described in Section 4.2 was applied and evaluated using simulated data (Figure 4), which was generated under the complex Wishart distribution. This simulation was based on nine classes observed in a SIR-C L-band PolSAR image: River, Caatinga, Prepared Soil, Soybean in three different phenological stages, Tillage, and Corn in two phenological stages. For each class, 150×150 pixels with 4-looks were simulated, totalizing an image of 450×450 pixels. The covariance matrices used in the simulation are given by:

$$\Sigma_{\text{River}} = \begin{bmatrix} 2.98 \cdot 10^{-3} & 5.31 \cdot 10^{-6} + \jmath\, 8.11 \cdot 10^{-5} & 3.47 \cdot 10^{-3} + \jmath\, 3.42 \cdot 10^{-4} \\ & 3.40 \cdot 10^{-4} & 4.47 \cdot 10^{-6} + \jmath\, 1.39 \cdot 10^{-4} \\ & & 1.19 \cdot 10^{-2} \end{bmatrix}$$

$$\Sigma_{\text{Caatinga}} = \begin{bmatrix} 1.11 \cdot 10^{-1} & -3.10 \cdot 10^{-3} - \jmath\, 1.58 \cdot 10^{-3} & 1.98 \cdot 10^{-2} + \jmath\, 1.65 \cdot 10^{-3} \\ & 3.40 \cdot 10^{-2} & -1.41 \cdot 10^{-3} + \jmath\, 1.87 \cdot 10^{-3} \\ & & 9.47 \cdot 10^{-2} \end{bmatrix}$$

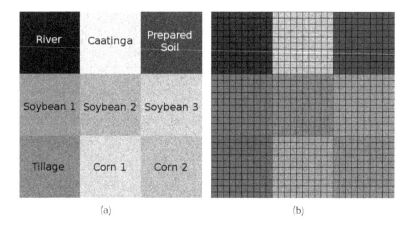

(a)   (b)

Figure 4: Simulated PolSAR image with 4-looks: (a) intensities color composition - HH(R), HV(G), VV(B) and (b) segmentation scheme in 15×15 pixels segments.

$$\Sigma_{\text{Prepared Soil}} = \begin{bmatrix} 1.05 \cdot 10^{-2} & -5.39 \cdot 10^{-6} - \jmath\, 2.37 \cdot 10^{-4} & 7.53 \cdot 10^{-3} + \jmath\, 1.75 \cdot 10^{-3} \\ & 8.46 \cdot 10^{-4} & -3.38 \cdot 10^{-5} + \jmath\, 1.32 \cdot 10^{-4} \\ & & 1.14 \cdot 10^{-2} \end{bmatrix}$$

$$\Sigma_{\text{Soybean 1}} = \begin{bmatrix} 3.40 \cdot 10^{-2} & -1.79 \cdot 10^{-3} - \jmath\, 1.86 \cdot 10^{-3} & -3.6 \cdot 10^{-4} - \jmath\, 7.58 \cdot 10^{-3} \\ & 5.16 \cdot 10^{-3} & 4.38 \cdot 10^{-4} + \jmath\, 4.28 \cdot 10^{-4} \\ & & 5.38 \cdot 10^{-2} \end{bmatrix}$$

$$\Sigma_{\text{Soybean 2}} = \begin{bmatrix} 4.31 \cdot 10^{-2} & -1.76 \cdot 10^{-3} - \jmath\, 1.32 \cdot 10^{-3} & -1.78 \cdot 10^{-4} - \jmath\, 1.73 \cdot 10^{-3} \\ & 9.26 \cdot 10^{-3} & 6.55 \cdot 10^{-4} + \jmath\, 1.27 \cdot 10^{-3} \\ & & 4.35 \cdot 10^{-2} \end{bmatrix}$$

$$\Sigma_{\text{Soybean 3}} = \begin{bmatrix} 7.53 \cdot 10^{-2} & -4.25 \cdot 10^{-3} - \jmath\, 7.66 \cdot 10^{-3} & 5.87 \cdot 10^{-4} - \jmath\, 1.36 \cdot 10^{-3} \\ & 1.47 \cdot 10^{-2} & -2.18 \cdot 10^{-4} + \jmath\, 1.21 \cdot 10^{-3} \\ & & 3.70 \cdot 10^{-2} \end{bmatrix}$$

$$\Sigma_{\text{Tillage}} = \begin{bmatrix} 3.53 \cdot 10^{-2} & 1.20 \cdot 10^{-3} + \jmath\, 1.02 \cdot 10^{-4} & 1.64 \cdot 10^{-2} - \jmath\, 2.65 \cdot 10^{-3} \\ & 3.05 \cdot 10^{-3} & 4.48 \cdot 10^{-4} + \jmath\, 1.88 \cdot 10^{-4} \\ & & 3.29 \cdot 10^{-2} \end{bmatrix}$$

$$\Sigma_{\text{Corn 1}} = \begin{bmatrix} 1.15 \cdot 10^{-1} & -3.95 \cdot 10^{-3} - \jmath\, 3.57 \cdot 10^{-3} & 9.13 \cdot 10^{-3} - \jmath\, 4.86 \cdot 10^{-3} \\ & 1.33 \cdot 10^{-2} & 3.34 \cdot 10^{-3} + \jmath\, 2.83 \cdot 10^{-3} \\ & & 1.47 \cdot 10^{-1} \end{bmatrix}$$

$$\Sigma_{\text{Corn 2}} = \begin{bmatrix} 4.19 \cdot 10^{-2} & 1.08 \cdot 10^{-3} - \jmath\, 1.01 \cdot 10^{-3} & 9.24 \cdot 10^{-3} - \jmath\, 3.68 \cdot 10^{-3} \\ & 1.02 \cdot 10^{-2} & 2.43 \cdot 10^{-4} + \jmath\, 3.31 \cdot 10^{-4} \\ & & 5.71 \cdot 10^{-2} \end{bmatrix}$$

The region classification procedure was applied using four different segmented images with segments of sizes 5×5, 10×10, 15×15 and 30×30 pixels, respectively. The 15×15 segmented image is presented in Figure 4(b). The prototype of each class, also needed for the classification procedure, was generated by sampling 900 pixels, representing a training sample of 30×30 pixels for each class. The simulation of the prototypes, performed independently of the simulated image, ensures that identical data are not being considered in the computation of test statistics and, consequently, in the determination of the corresponding p-values.

The results presented in Figure 5 and in Table 3 by Silva et al. (2013) are compatible with the theoretically expected values. The hypothesis tests rejection rates were approximately 5% for all segmentation cases and stochastic distances, except when the χ^2 distance was used, and the Bhattacharyya Gaussian distance was applied to small (5×5 pixels) segments. The rejection rates for the χ^2 distance were higher than the theoretical values in all segmentation cases, reaching the value of approximately 24.5% for the segmentation of 5×5 pixels segments. The poor performance of the χ^2 distance test statistic was also observed by Frery et al. (2011).

Table 3: Percentage of segments for which H_0 was not rejected at 5% significance level, for simulated data case.

Distances	Percentage (%)			
	5×5 pixels 8100 segments	10×10 pixels 2025 segments	15×15 pixels 900 segments	30×30 pixels 225 segments
Bhattacharyya	94.0	95.2	94.3	93.8
Kullback-Leibler	93.7	95.1	94.3	93.3
Hellinger	95.2	95.3	94.8	93.8
Rényi (order $\beta = 0.9$)	93.8	95.1	94.3	93.8
χ^2	75.5	91.2	92.8	92.4
Bhattacharyya Gaussian	90.6	94.1	95.1	98.2

6   Final considerations

In this article two different Polarimetric SAR image processing tools using stochastic distances were presented. These tools consist of a filter and of a digital image classifier.

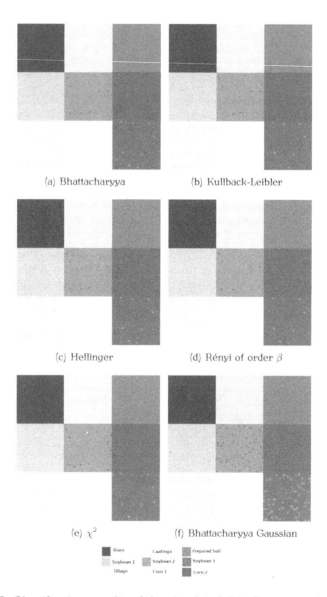

(a) Bhattacharyya (b) Kullback-Leibler

(c) Hellinger (d) Rényi of order β

(e) χ^2 (f) Bhattacharyya Gaussian

Figure 5: Classification results of the simulated data for segments of size 5×5 pixels.

The filter belongs to the Nonlocal Means family filters, and the free-noise pixels are obtained from a weighted average of selected pixels from statistical tests derived from stochastic distances.

The developed classifier is a region classifier that adopts the supervised classification method. The regions of the image are assigned to the previously determined classes by the evaluation of the statistical tests based on stochastic distances. An important feature of this classifier is a generation of an uncertainty map associated to the classified image.

The applications of the developed PolSAR image processing tools showed results in accordance to and better than those presented in the literature. Therefore, the use of stochastic distances derived from an appropriate statistical modeling, and its associated statistical tests, is highly relevant and promising in the context of polarimetric radar images processing.

References

Buades, A., Coll, B. & Morel, J. (2005), 'A review of image denoising algorithms, with a new one', *Multiscale Modeling & Simulation* 4(2), 490–530. DOI 10.1137/040616024.

Burbea, J. & Rao, C. (1982), 'Entropy differential metric, distance and divergence measures in probability spaces: a unified approach', *Journal of Multivariate Analysis* 12, 575–596.

Chen, J., Chen, Y., An, W., Cui, Y. & Yang, J. (2011), 'Nonlocal filtering for Polarimetric SAR data: A pretest approach', *IEEE Transactions on Geoscience and Remote Sensing* 49(5), 1744–1754. DOI 10.1109/TGRS.2010.2087763.

Conradsen, K., Nielsen, A. A., Schou, J. & Skriver, H. (2003), 'A test statistic in the complex Wishart distribution and its application to change detection in polarimetric SAR data', *IEEE Transactions on Geoscience and Remote Sensing* 41(1), 4–19. DOI 10.1109/TGRS.2002.808066.

Deledalle, C.-A., Denis, L., Tupin, F., Reigber, A. & Jäger, M. (2013), NL-SAR: A unified non-local framework for resolution-preserving (Pol)(In)SAR denoising, Tech Report 1, Institut de Mathématiques de Bordeaux (IMB), Talence, France.

Deledalle, C.-A., Tupin, F. & Denis, L. (2010), Polarimetric SAR estimation based on non-local means, *in* 'IEEE International Geoscience and Remote Sensing Symposium (IGARSS)', Honolulu, Hawaii. DOI 10.1109/IGARSS.2010.5653936.

Frery, A. C., Müller, H.-J., Yanasse, C. C. F. & Sant'Anna, S. J. S. (1997), 'A model for extremely heterogeneous clutter', *IEEE Transactions on Geoscience and Remote Sensing* **35**(3), 648–659. DOI 10.1109/36.581981.

Frery, A. C., Nascimento, A. D. C. & Cintra, R. J. (2011), 'Information theory and image understanding: An application to polarimetric SAR imagery', *Chilean Journal of Statistics* **2**(2), 81–100.

Frery, A. C., Nascimento, A. D. C. & Cintra, R. J. (2014), 'Analytic expressions for stochastic distances between relaxed complex Wishart distributions', *IEEE Transactions on Geoscience and Remote Sensing* **52**(2), 1213–1226. DOI 10.1109/TGRS.2013.2248737.

Gao, G. (2010), 'Statistical modeling of SAR images: A Survey', *Sensors* **10**(1), 775–795. DOI 10.3390/s100100775.

Goodman, J. W. (1976), 'Some fundamental properties of speckle', *Journal of the Optical Society of America* **66**(11), 1145–1150. DOI 10.1364/JOSA.66.001145.

Goodman, N. R. (1963), 'Statistical analysis based on a certatin Multivariate Complex Gaussian distribution (An introduction)', *The Annals of Mathematical Statistics* **34**(1), 152–177. DOI 10.1214/aoms/1177704250.

Goudail, F. & Réfrégier, P. (2004), 'Contrast definition for optical coherent polarimetric images', *IEEE Transactions on Pattern Analysis and Machine Intelligence* **26**(7), 947–951. DOI 10.1109/TPAMI.2004.22.

Jager, L. & Wellner, J. (2007), 'Goodness-of-fit tests via phi-divergences', *The Annals of Statistics* **35**(5), 2018–2053. DOI 10.1214/0009053607000000244.

Kailath, T. (1967), 'The divergence and Bhattacharyya distance measures in signal selection', *IEEE Transactions on Communication Technology* **15**(1), 52–60. DOI 10.1109/TCOM.1967.1089532.

Lee, J.-S., Grunes, M. R. & de Grandi, G. (1999), 'Polarimetric SAR speckle filtering and its implication for classification', *IEEE Transactions on Geoscience and Remote Sensing* **37**(5), 2363–2373. DOI 10.1109/36.789635.

Lee, J.-S., Grunes, M. R., Schuler, D. L., Pottier, E. & Ferro-Famil, L. (2006), 'Scattering-model-based speckle filtering of polarimetric SAR data', *IEEE Transactions on Geoscience and Remote Sensing* **44**(1), 176–187. DOI 10.1109/TGRS.2005.859338.

Lee, J. S., Hoppel, K. W., Mango, S. A. & Miller, A. R. (1994), 'Intensity and phase statistics of multilook polarimetric and interferometric SAR imagery', *IEEE Transactions on Geoscience and Remote Sensing* **32**(5), 1017–1028. DOI 10.1109/36.312890. ISSN 0196-2892.

Mittal, A., Moorthy, A. K. & Bovik, A. C. (2012), 'No-reference image quality assessment in the spatial domain', *IEEE Transactions on Image Processing* **21**(12), 4695–4708. DOI 10.1109/TIP.2012.2214050.

Nascimento, A. D. C. (2012), Teoria Estatística da Informação para Dados de Radar de Abertura Sintética Univariados e Polarimétricos, PhD Thesis, Universidade Federal de Pernambuco, Recife.

Nascimento, A. D. C., Cintra, R. J. & Frery, A. C. (2010), 'Hypothesis testing in speckled data with stochastic distances', *IEEE Transactions on Geoscience and Remote Sensing* **48**(1), 373–385. DOI 10.1109/TGRS.2009.2025498.

Salicrú, M., Menéndez, M. L., Pardo, L. & Morales, D. (1994), 'On the applications of divergence type measures in testing statistical hypothesis', *Journal of Multivariate Analysis* **51**(2), 372–391. DOI 10.1006/jmva.1994.1068.

Seghouane, A. K. & Amari, S. I. (2007), 'The AIC criterion and symmetrizing the Kullback-Leibler divergence', *IEEE Transactions on Neural Networks* **18**(1), 97–106. DOI 10.1109/TNN.2006.882813.

Silva, W. B. d. (2013), Classificação de regiões de imagens utilizando testes de hipótese baseados em distâncias estocásticas: aplicações a dados polarimétricos, PhD thesis, Instituto Nacional de Pesquisas Espaciais (INPE), São José dos Campos. URL http://urlib.net/sid.inpe.br/mtc-m19/2013/02.20.16.22.

Silva, W. B., Freitas, C. C., Sant'Anna, S. J. S. & Frery, A. C. (2013), 'Classification of segments in PolSAR imagery by minimum stochastic distances between wishart distributions', *IEEE Journal of Selected Topics in Applied Earth Observations and Remote Sensing* **6**(3), 1263–1273. DOI 10.1109/JSTARS.2013.2248132.

Theodoridis, S. & Koutroumbas, K. (2008), *Pattern Recognition*, 4th ed., Academic Press, San Diego, CA. ISBN: 978-1597492720.

Torres, L., Sant'Anna, S. J. S., Freitas, C. C. & Frery, A. C. (2014), 'Speckle reduction in polarimetric SAR imagery with stochastic distances and nonlocal means', *Pattern Recognition* **47**(1), 141–157. DOI 10.1016/j.patcog.2013.04.001.

Ulaby, F. T. & Elachi, C. (1990), *Radar Polarimetriy for Geoscience Applications*, Artech House, Norwood. ISBN: 978-0890064061.

Vasile, G., Trouve, E., Lee, J.-S. & Buzuloiu, V. (2006), 'Intensity-driven adaptive-neighborhood technique for Polarimetric and Interferometric SAR parameters estimation', *IEEE Transactions on Geoscience and Remote Sensing* **44**(6), 1609–1621. DOI 10.1109/TGRS.2005.864142.

Zhong, H., Zhang, J. & Liu, G. (2014), 'Robust polarimetric SAR despeckling based on nonlocal means and distributed Lee filter', *IEEE Transactions on Geoscience and Remote Sensing* **52**(7), 1–13. DOI 10.1109/TGRS.2013.2280278.

Semi-supervised learning applied to multilook PolSAR imagery[‡]

Michelle M. Horta*   Nelson D. A. Mascarenhas*

Alejandro C. Frery[†]

* Departamento de Computação
Universidade Federal de São Carlos
michellemh@gmail.com, nelson@dc.ufscar.br

† Instituto de Computação
Universidade Federal de Alagoas
acfrery@pesquisador.cnpq.br

1 Introduction

Polarimetric SAR images are built from polarimetric radar returns. Some SAR image advantages are the independence from cloud cover and solar illumination conditions. However, the presence of multiplicative speckle noises degrades the processing of this type of image (Frery et al., 2007; Cao et al., 2007).

The multilook PolSAR data are hermitian matrices that have been classically described by the Wishart distribution (Cao et al., 2007). Despite its simplicity, the Wishart distribution was proposed to model targets with more homogeneous features, as pasture areas. In Frery et al. (2007), the $\mathcal{G}_P^0$ distribution was proposed to model targets with variability. The $\mathcal{G}_P^0$ law is a flexible model and, depending on the conditions, it can have the Wishart distribution as a particular case.

Many multilook PolSAR applications researches have emerged using the classical types of learning: supervised and unsupervised. However, the semi-supervised learning can become a promising research topic (Frery et al., 2007; Cao et al., 2007; Horta & Mascarenhas, 2010). In Hänsch & Hellwich (2009), a semi-supervised classification was applied to multilook PolSAR images using a deterministic annealing clustering combined with a multi-layer perceptron supervised classifier.

‡This work was supported by FAPESP under grant number 2009/14270-4

This paper tests if the labeled data can increase the performance of three statistical clustering techniques when they are adapted to a semi-supervised learning: two of them use the stochastic expectation-maximization algorithm with the mixture of Wishart or $\mathcal{G}_P^0$ distributions, as shown in Horta & Mascarenhas (2010); the third method, proposed in Cao et al. (2007), combines the k-means algorithm with the Wishart distribution.

The remainder of this paper is organized as follows: Section 2 shows the multilook PolSAR models, Sections 3 and 4 describes the classification methods, Sections 5 and 6 present the experimental results and conclusions, respectively.

2   Models for fully PolSAR data

PolSAR images incorporate the scattering properties of the target due to the combination of transmit and receive polarizations. Representing the mono-static reciprocal scattering of single-look vector by $s = (S_{hh}, S_{hv}, S_{vv})$, $S_{pq} \in \mathbb{C}$, the multilook PolSAR data is a hermitian matrix given by:

$$Z = \frac{1}{n} \sum_{i=1}^{n} s_i s_i^{*t} = \begin{pmatrix} Z_{hh} & Z_{hhhv} & Z_{hhvv} \\ Z_{hhhv}^* & Z_{hv} & Z_{hvvv} \\ Z_{hhvv}^* & Z_{hvvv}^* & Z_{vv} \end{pmatrix}, \tag{1}$$

where $*$ is the conjugate, t is the transpose and n is the nominal number of looks. The diagonal elements are the multilook intensities of the PolSAR data. The multilook PolSAR data is known as n−look covariance matrix (Frery et al., 2007; Cao et al., 2007). In this paper, the image data are realizations of independent and identically distributed random variables that can obey Wishart distributions or $\mathcal{G}_P^0$ distributions. The Wishart distribution describes homogeneous areas as river and pasture, i.e., targets keeping minimum variability (Frery et al., 2007; Cao et al., 2007). The density is defined by:

$$f(z; C, n) = \frac{n^{mn} |z|^{n-m}}{|C|^n h(m,n)} \exp\left\{ -n \operatorname{tr}\left(C^{-1} z \right) \right\}, \tag{2}$$

where $h(m,n) = \pi^{m(m-1)/2} \Gamma(n) \dots \Gamma(n - m + 1), n \geq m$, n is the number of looks, $m = 3$ is the number of polarizations, tr and $|\cdot|$ are the trace and the determinant, respectively. The parameter C is the expected value of the random matrix Z.

The $\mathcal{G}_P^0$ distribution describes heterogeneous areas like forest and urban regions. The variability of these textured areas is supported by the roughness parameter α. When $\alpha \to -\infty$, the $\mathcal{G}_P^0$ law features homogeneous areas, and when $\alpha \to -1$, the $\mathcal{G}_P^0$ law describes heterogeneous areas (Frery et al., 2007). The density is given by:

$$f\left(z; \alpha, C, n\right) = \frac{n^{mn} \left|z\right|^{n-m} \Gamma\left(mn - \alpha\right)}{\left|C\right|^n h(m,n) \Gamma\left(-\alpha\right) \left(-\alpha - 1\right)^\alpha} \left(n \operatorname{tr}\left(C^{-1} z\right) + \left(-\alpha - 1\right)\right)^{\alpha - mn}.$$

(3)

2.1 Estimators by moments method

In Frery et al. (2007), the parameters of the Wishart and $\mathcal{G}_P^0$ distributions are defined by the moments estimators. Then, the sample n-look covariance matrix is the first order sample moment $\widehat{C} = \widehat{m}_1(\tilde{Z})$, with sample $\tilde{Z} = (Z_1, \ldots, Z_N)$. Whereas the roughness estimator is computed by the average of the estimates $\widehat{\alpha}_\ell$ of each intensity datum Z_ℓ, $\ell \in \{hh, hv, vv\}$ given by:

$$\Gamma(\widehat{\alpha}_\ell, n) - (\widehat{m}_{1/4}^2(\tilde{Z}_\ell) / \widehat{m}_{1/2}(\tilde{Z}_\ell)) = 0,$$

where

$$\Gamma\left(\widehat{\alpha}_\ell, n\right) = \frac{\Gamma^2\left(-\widehat{\alpha}_\ell - 1/4\right) \Gamma^2\left(n + 1/4\right)}{\Gamma\left(-\widehat{\alpha}_\ell - 1/2\right) \Gamma\left(n + 1/2\right) \Gamma\left(-\widehat{\alpha}_\ell\right) \Gamma\left(n\right)}.$$

3 Classification method

Three clustering algorithms are applied to semi-supervised classifications for comparison analysis: SEM algorithm using the mixture of Wishart distribution, SEM algorithm using the mixture of $\mathcal{G}_P^0$ distribution, and the k-means based on the Wishart distribution.

The former clustering algorithm is shown in Horta & Mascarenhas (2010). It combines the SEM algorithm with moments estimators in order to fit the Wishart mixture models. The SEM algorithm is a stochastic version of the classical expectation-maximization (EM) method (Côme et al., 2009; Baraldi et al., 2006), which is summarized as follows:

1. E-STEP: Update the posterior probabilities for each pixel and distribution by

$$\tau_{ij}^{(k)} = \frac{\rho_j^{(k)} f(z_i, \theta_j^{(k)})}{\sum_{\ell=1}^g \rho_\ell^{(k)} f(z_i, \theta_\ell^{(k)})},$$

using the density (2) or (3) according to the used distribution (Wishart or $\mathcal{G}_P^0$);

2. S-STEP: Randomly sample a label for each ith pixel according to the current estimated $\tau_{ij}^{(k)}$ value. Therefore, the image is partitioned in g classes represented as $\{Q_1^{(k)}, \ldots, Q_g^{(k)}\}$;

3. M-STEP: Update the parameter estimates by $\widehat{\rho}_j^{(k+1)} = \widehat{Q}_j^{(k)}/N$ and $\widehat{\theta}_j^{(k+1)}$ with the estimators of Section 2.1 and the pixels within the class $Q_j^{(k)}$.

The algorithm alternates between those three steps until convergence is achieved. The image can be classified using a maximum a posteriori decision rule.

The k-means algorithm is the complex Wishart clustering described in Cao et al. (2007) and is defined as follows:

1. STEP 1: Generate the image partition in g classes $\left\{Q_1^{(k)}, \ldots, Q_g^{(k)}\right\}$ and using the minimum distance

$$d\left(Z_i, C_j^{(k)}\right) = \ln |C_j| + \text{tr}(C_j^{-1} Z_i);$$

2. STEP 2: Calculate the centroids $\hat{C}_j^{(k+1)}, 1 \leq j \leq g$ using the moment estimator (Section 2.1) and the pixels within the class $Q_j^{(k)}$.

The algorithm alternates between those two steps until convergence is achieved. The image can be classified using the minimum distance.

In all cases, the initial parameters can be randomly initialized.

4 Semi-supervised classification method

In many remote sensing applications, although the labeled data are available, there are some critical problems, as the small sample size and the unrepresentative sample (Baraldi et al., 2006). On the other hand, the unsupervised processing can provide misunderstood results. Therefore, this paper analyzes the multilook PolSAR image clustering problems as a semi-supervised learning approach. In Horta & Mascarenhas (2010); Hänsch & Hellwich (2009), there are some previous studies about this problem.

Assuming that $Z^{SS} = [(\mathbf{Z}_1, c_1), \ldots, (\mathbf{Z}_m, c_m), \mathbf{Z}_{m+1}, \ldots, \mathbf{Z}_N]$ is the set comprising labeled data $Z^L = [(\mathbf{Z}_1, c_1), \ldots, (\mathbf{Z}_m, c_m)]$ and unlabeled data $Z^U = [\mathbf{Z}_{m+1}, \ldots, \mathbf{Z}_N]$, the methods can be adapted from two approaches: the seed and a semi-supervised one.

The seed approach treats the labeled data only in the initialization step. Then, in the clustering algorithms (Section 3), the initial parameters are calculated using the labeled data Z^L, while the SEM and k-mean algorithms are performed according to the unlabeled data Z^U.

The semi-supervised approach considers the labeled data in all steps of the clustering process, i.e., the labeled data will influence all clustering iterations. Hence, some modifications are defined as follows:

1. Calculate the initial parameters using the labeled data Z^L;

2. Clustering algorithms:

    (a) Update the proximity measures of the unlabeled data Z^U, respectively. In the SEM algorithm, update the posterior probabilities (E-STEP). In the k-means algorithm, update the distances of the first step;

    (b) Update the partitions according to the new proximity measures of the unlabeled data Z^U and preserve classes of the labeled data Z^L. In the SEM algorithm, it represents the S-STEP. In the k-means algorithm, it occurs in the first step;

    (c) Update the parameter estimates using the complete data set Z^{SS}. In the SEM algorithm, it represents the M-STEP. In the k-means algorithm, it represents the second step.

5 Results

The semi-supervised classification was applied to a synthetic and a real multilook PolSAR image. The three statistical clustering algorithms were analyzed considering two semi-supervised approaches: the labeled data are the seeds (seed approach) or the labeled data are included in the clustering algorithms (semi-supervised approach).

5.1 Synthetic image

The synthetic image has six regions as shown in Figure 1. Each region was randomly generated using $\mathcal{G}_P^0$ distribution with parameter set $(\mathrm{tr}(\hat{C}), \hat{\alpha}, \hat{n} = 4)$, respectively: region R1 - $\mathcal{G}_P^0(31, -20)$, region R2 - $\mathcal{G}_P^0(161, -10)$, region R3 - $\mathcal{G}_P^0(1256, -6.5)$, region R4 - $\mathcal{G}_P^0(1988, -4)$, region R5 - $\mathcal{G}_P^0(4056, -1.2)$ and region R6 - $\mathcal{G}_P^0(13604, -2.5)$. Then, regions R1 and R2 represent homogeneous areas, while regions R3, R4 and R5 describe heterogeneous areas.

Figure 1 also depicts the five analyzed labeled data, where each set is represented by a color. The green sample has 20% of the image. In this context, the blue, yellow and orange areas show labeled data with 10%, 1% and 0.5% of the image, respectively. The red areas represent a noisy labeled data in order to study the influence of wrong labeled samples.

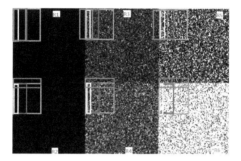

Figure 1: Synthetic image (200×300 pixels) and five labeled data sets represented by colors.

The classification results of the synthetic image are shown separately. Figure 2 depicts the performance of the semi-supervised SEM algorithm with $\mathcal{G}_P^0$ distribution (SEM- $\mathcal{G}_P^0$), while Figure 3 shows the results obtained with Wishart distributions and SEM (SEM-Wishart) or k-means algorithms. All results were obtained at iteration 20.

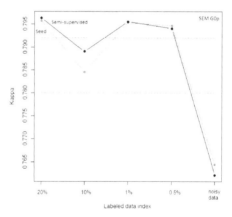

Figure 2: Kappa coefficient of agreement of the semi-supervised SEM classifications with $\mathcal{G}_P^0$ law. Red strip delimit the results that are not statistically significant according to kappa and its variance.

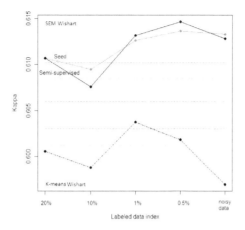

Figure 3: Kappa coefficient of agreement of the semi-supervised classifications with the Wishart law. The colour strips delimit the results that are not statistically significant according to kappa and its variance.

The SEM-$\mathcal{G}_P^0$ algorithm provided the best results, even with the decline observed with the noisy labeled data. By analyzing each clustering method individually, there were no differences between the results provided by the two semi-supervised approaches, specially, with the k-means results.

5.2   Real image

Figure 4 shows a small area of the classical image of San Francisco, CA, obtained by the AIRSAR sensor with the L-band and 4 nominal number of looks. It consists of three main classes: sea, vegetation and urban areas. The labeled data are represented by green, blue and red samples, where the red one defines the noisy labeled data. The green labeled data represents 12% of the image, the blue data consists of 6% and the red sample has 8%.

The classifications are also obtained at iteration 20 and the kappa coefficient results are shown in Figure 5. The results are not statistically significant for the same clustering method even with different types of labeled data or semi-supervised approaches. The differences occur between the clustering methods, where the SEM with $\mathcal{G}_P^0$ distributions performed the best results.

Figure 4: Small area (140×140 pixels) of the San Francisco PolSAR image.

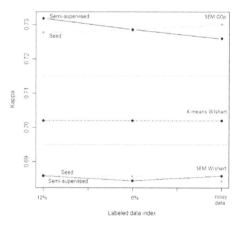

Figure 5: Kappa coefficient of agreement of the semi-supervised classifications. The red strips delimit the results that are not statistically significant according to kappa and its variance.

6 Conclusions

This paper analyzes the inclusion of sets with few labeled data in the multilook PolSAR image classification using clustering of $\mathcal{G}_P^0$ or Wishart distributions. The synthetic tests showed that the semi-supervised clustering with $\mathcal{G}_P^0$ distributions can be affected by labeled data with noise, while the methods based on the Wishart distributions are less influenced. However the SEM with $\mathcal{G}_P^0$ distributions provided the best kappa coefficient results, even with the noisy labeled data. The classifications with the real image strengthen this idea, showing that the semi-supervised adaptation of the clustering analyzed algorithms can be robust even with noisy labeled data.

References

Baraldi, A., Bruzzone, L. & Blonda, P. (2006), 'A multiscale expectation-maximization semi-supervised classifier suitable for badly posed image classification', *IEEE Trans. on Image Processing* **15**(8), 2208–2225.

Cao, F., Hong, W., Wu, Y. & Pottier, E. (2007), 'An unsupervised segmentation with an adaptive number of clusters using the SPAN/H/α/A space and the complex Wishart clustering for fully polarimetric SAR data analysis', *IEEE Trans. on Geoscience and Remote Sensing* **45**(11), 3454–3467.

Côme, E., Oukhellou, L., Denoeux, T. & Aknin, P. (2009), 'Learning from partially supervised data using mixture models and belief functions', *Pattern Recognition* **42**(3), 334–348.

Frery, A. C., Correia, A. H. & Freitas, C. C. (2007), 'Classifying multifrequency fully polarimetric imagery with multiple sources of statistical evidence and contextual information', *IEEE Trans. on Geoscience and Remote Sensing* **45**(10), 3098–3109.

Hänsch, R. & Hellwich, O. (2009), Semi-supervised learning for classification of polarimetric SAR data, *in* 'Proceedings of the IEEE International Geoscience and Remote Sensing Symposium', pp. III–987–III–990.

Horta, M. M. & Mascarenhas, N. D. A. (2010), Fully polarimetric sar image classification using different learning approaches, *in* 'Proceedings of the 17th International Conference on Systems, Signals and Image Processing', pp. 316–319.

Revisiting landscape views in information visualization[‡]

Pedro Henrique Siqueira[*]   Guilherme P. Telles[*]

Rosane Minghim[†]

[*] Instituto de Computação
Universidade Estadual de Campinas
opedrosiqueira@gmail.com, gpt@ic.unicamp.br

[†] Instituto de Ciências Matemáticas e de Computação
Universidade de São Paulo
rminghim@icmc.usp.br

1 Introduction

Information visualization concerns building visual layouts of abstract data objects to unveil relationships and patterns within a data set. In contrast with 3D object visualization (such as 3D for medical imagery and support for engineering), where 3D is a natural way to show target objects, abstract data have to be translated to a conceptual visual model before being presented to analysts.

For abstract data, many types of visual layouts have already been proposed that make use of both 2D and 3D views. Many of these tools, such as multidimensional projections, are capable of mapping data expressed as samples in a multidimensional space into visual spaces (2D or 3D) via transformations aimed at preserving some property (such as distances or neighborhoods). We focus here on visually enriching point placements, that is, in defining a surface view as output of techniques that produce sets of points placed on a 2D plane. We are concerned with adding information to that layout in order to improve its expressive power and allow additional analysis to be done by means of a single view.

On top of a point based presentation, visual clues can be added to increase analysis power, such as color, texture, or iconographic displays

[‡]PHS acknowledges the financial support of FAPESP (fellowship 2012/24634-6). GPT and RM acknowledge the financial support of FAPESP (grant 2011/22749-8) and CNPq.

for the points. A surface model can also be built by using the third dimension to reflect a particular variable or concept. That surface can reflect, through its shape, a global view of the data set regarding the relationship between the distance distribution of the points and the particular concept mapped to the third coordinate. The visualization of that model is known as a landscape view or information landscape. Landscape views have been proposed many years ago, and some have been implemented in software tools that are currently in use, such as In-spire (http://in-spire.pnnl.gov/about.stm), a text visualization software. One of the problems with most currently available landscape views for abstract data is that the basis for landscape construction is subject to the uncertainty of the point placement strategy.

In point placements, distances between dots in 2D or 3D visual space are meant to reflect distance between data objects in original multidimensional space. Another spatial representation may be obtained by coloring the space among dots, producing a kind of heatmap that is a flat landscape. Variation in height associated with points or regions produce a surface embedded in 3D, with an additional degree of information given by the shapes formed in surfaces regions. Including visual artifacts such as color and icon shapes can assist exploration in both dot and landscape displays.

The effectiveness of any 3D views for data exploration has been widely debated. Advantages and disadvantages have been observed as will be recalled later. While the debate is still open, it is also illustrated by many recurring strategies that a good visual design and a simple mapping to the third dimension can help adding expression where a 2D view is already cluttered or limited to convey, at once, general and particular views of the data set. This last case, that is, adding expressive power for better exploration, is the most central point in this work.

The mappings shown in this work are actually on a surface, which for most purposes is a 2D element, however embedded in 3D. Visualizations of the resulting object may suffer from some of the limitations in 3D visualizations, such as the effects of occlusion and difficulties interacting with the visualization. The input to the visualization, that is, the 2D projection, is done employing later techniques, capable of high degree of precision in terms of point neighborhoods. Also, the produced landscapes are enriched by visual mappings that support summarization and have being employed successfully in various applications.

The result of this effort is a revisited and improved landscape for visualizing and exploring a set of data objects. We present a pipeline for the construction of the surface landscapes for multidimensional data points, supported by newer projection techniques and visually improved by segmentation algorithms, guiding lines and textures connected to the data, as well as by a 2D cursor for interaction. The result is amenable to user exploration, and reinforces our belief that geographic metaphors can be an effective exploratory environment for data objects, taking advantage of the third dimension while keeping users mental references stable.

The rest of this article is organized as follows. In Section 2, we recall the related work on landscape views and we group the issues surrounding the 2D/3D question. In Section 3, we present our pipeline, together with examples of landscapes for non-hierarchical data. In Section 4, we present our concluding remarks.

2   Related work

The geographic metaphor, where data which is not geographical in nature are represented as a landscape, is prevalent in information visualization. The intention is that the user should benefit from the familiar notion of a map and could then explore data objects guided by neighboring relations intrinsically imposed by geometry and represented by mountains and valleys. Fabrikant et al. (2010) suggests that landscape layouts work not because people understand geomorphological landscapes but because "everyday experience with manipulable tabletop spaces is extended metaphorically to a wide variety of domains of other spatial scales", and then people associate higher with more and bigger with more.

Chalmers (1993) proposed reducing the dimensionality of a space obtained from the weighted term-frequency of a document collection, obtaining a 3D layout to which a horizon is added. The resulting layout resembles a map where documents are shown as icons. He points out advantages of the approach, such as the wide overview provided on the data, the representation continuity, and the preservation of the mental model. A remarkable point of that article is the reference to the book by Lynch (1960), that enumerates elements in a city map which are strongly related to the ability of building a city image and that were identified by persons during surveys. Such elements are paths, edges (linear elements that are not paths), districts (medium to large portions of the city), nodes (strategic points which the observer may enter while traversing the city), and landmarks (reference points). Lynch remarks that continuity and contrast should coexist and balance, and that a city view should be stable over time. He also states that such elements are present in models of more general environments, not only in cities models.

Wise et al. (1995) introduced ThemeScape, that starts from a collection of texts, builds a vector representation and then applies a multidimensional reduction to obtain a 2D space. Theme strength of each point is mapped to the third dimension to produce a landscape. The authors highlight that the layout is suitable for revealing interrelationships among documents in either large or small scale. Using a multidimensional reduction to obtain a map is a prevalent procedure, although computationally expensive in the past.

There are many multidimensional reduction techniques in the literature, such as principal component analysis (Jolliffe, 2002), self-organizing maps (Kohonen et al., 2001) force-directed placement (Fruchterman & Reingold, 1991), and multidimensional projections. More recently a large improvement was made on techniques for multidimensional projections resulting in algorithms that are general, fast and precise. This improves on previous algorithms that were either slow, subject to large deviations regarding the original space or restricted to data having specific patterns. Among these techniques we cite ProjClus (Paulovich & Minghim, 2006) and Least-Square Projection (LSP – Paulovich et al., 2006).

Skupin & Buttenfield (1996) proposed building landscapes for newspaper articles applying MDS to a binary vector space of term presence in articles, then mapping the 3rd dimension to the number of terms in each article. They report using a geographic information system for rendering and providing interaction with the display, both in 2D and in 3D.

Davidson et al. (1998) introduced a system (VxInsight) for information visualization that represents data and their relationships as a graph and them produces displays through Laplacian eigenvectors or force-based placement of vertices. A landscape is built mapping density to the 3rd dimension. The display may be augmented by showing edges between objects and by labeling.

Boyack et al. (2002) build a 2D space from scientific articles applying a force-directed scheme to text attributes, and then add a third dimension that reflects density, obtaining a landscape. They also allow adding arrows between data points to indicate citations, augmenting the display with different symbols.

Bischoff et al. (2004) suggest combining ThemeScape with the ThemeRiver (Havre et al., 2002) to analyze a document collection over time. They also superimpose a tree on the landscape to highlight cross-references among documents.

Skupin (2004) builds a multi-level 2D landscape for a set of scientific abstracts in cartography. The lowest zoom level has the documents represented as points. The next zoom level groups points related by the occurrence of specific terms into clusters. Each subsequent zoom level groups clusters in the previous one under more general terms. The highest zoom level depicts large clusters that represent broad terms in the area. Hall & Clough (2013) build a similar visualization.

Jaffe et al. (2006) introduced Tag Maps, that superimpose a word-cloud on a map. Their application used word-clouds built from tags added to photos that were geo-referenced. They report that users were positive regarding Tag Maps for summarizing a map region.

In a comprehensive study on points and landscapes, Tory et al. (2007) compared seven different dot and landscape representations of data: colored and grayscale points; colored and grayscale 2D landscapes; colored, grayscale and uncolored 3D landscapes. Users were evaluated in the task of selecting a display region that contained the most points of a specified target value range, which were most often represented by a single color or gray value. Landscape displays varied in visual complexity (regarding levels) along the experiments. Users were asked to rate the displays with respect to the task they performed. The authors concluded that colored points were the most accurate, fast and highly rated display. Landscapes were from 4 to 10% less accurate and from 1.9 to 4.2 times slower than points. They concluded that 2D was faster than 3D landscapes with no difference in accuracy, and hypothesize that occlusion and the need to rotate the display may have contributed to the difference. They also reported that color displays were faster than grayscale displays, and that uncolored 3D displays were the worst.

In a posterior experiment, Tory et al. (2009) found that people memorized dot displays more accurately than 2D or 3D landscapes, hypothesizing that the extra features of landscapes may be distracting. 3D landscapes were found to be more accurately memorized than 2D landscapes. They also reported that in most cases the memory accuracy increased with points density.

In the work of Jianu & Laidlaw (2013), a set of genes is represented on map such that the distance is proportional to the dissimilarity of expression profiles under multiple biological conditions. The interaction with the map is done though zoom and pan. Different levels are represented with a different layout, for instance, point, glyphs and heatmaps.

Similar to the landscape metaphor is the islands metaphor, that includes water and islands to the map, and was used by Pampalk (2001) to visualize music datasets.

Much work in visualization points to the differences among 2D, 3D and combined displays, regarding both the user ability to interact with the display and user's performance on exploratory tasks. There seems to be an undisputed matter the conclusion that 3D displays, or the combination of 3D and 2D displays, are superior when the task is to visualize models of real world objects, scenes or phenomena.

In other contexts, including in information visualization, the use of 3D has received critics that range from mild to acute. The negative characters of 3D layouts that are most frequently pointed out are the occlusion of objects, and the difficulty to maintain a reference with respect to the viewing angle and to the space axes, impairing navigation and orientation. Dealing with such characters demands, for instance, removing layers, using transparency or providing additional views, at the cost of visual stability and context loss, and additional complexity to the user.

For instance, Cockburn & McKenzie (2001) have noticed no improvement of 3D over 2D data mountains (Robertson et al., 1998) in tasks of organizing and retrieving web pages represented as thumbnails. Westerman & Cribbin (2000) concluded that 2D is more effective than 3D for searches in a virtual space built from data, but may be better for browsing and to visualize complex relationships among data.

Risden et al. (2000) compared the performance of specialized users in searching and updating categories of web-page indices both presented as lists and sublists in 2D and as 3D graphs in a hyperbolic display. They observed that 3D improved speed while preserved precision in search tasks, but 2D was more effective for the creation of new categories.

Piringer et al. (2004) analyzed scatter-plots in 2D and in 3D, and remark the difficulty of recognizing point density and depth in 3D views. They propose alternative representations for the points to improve point discrimination, but note that for dense spaces many data objects may be represented by a single pixel, which is a limitation that is difficult to circumvent. Tory et al. (2006) analyzed 2D, 3D and combined displays for estimating relative position, orientation and volume of objects, and conclude that 3D is suitable for approximated navigation and positioning, but are not precise in general for precise positioning and navigation. In his book Mazza (2009) discourages the use of 3D except for real-world visualizations.

Fabrikant et al. (2014) conducted a study with network spatializations and found no improvement in the ability of users to identify similar documents in 3D compared to 2D, hypothesizing that the extra information provided by 3D may not be worth the extra cognitive effort.

Many of the problems encountered with 3D user studies refer to lack of reference to support recognizing the patterns in 3D with the same models as in 2D, and with difficulties in interaction. However, along the last years there were very informative and successful visualizations that employed the third dimension to allow users to interpret additional information or to reduce information loss due to mapping from higher dimensions.

One such example shows the use of 2.5 visualizations to explore biological networks (Fung et al., 2008). While graphs in general are particularly difficult to understand from a 3D representation, a possible 2.5D alternative is to draw parts of the graph in layers, offering thus references for recognition. The article shows a possible rendering for a network with various 'piled up' slices, each one with a sector of the graph that is connected to the other slices or planes through graph edges.

Another example of 3D representation shows a mapping from multidimensional scaling algorithms onto 3D space rather than the usual 2D space (Nam & Mueller, 2013). The resulting 3D point cloud is difficult to interpret statically, but is incremented by the possibility of rotation, counting on motion parallax to allow the user to interpret proper clusters in the data. Interaction can be done over any 2D view from the 3D plot. Another work shows an alternative way to interact with 3D clouds, via clustering, which eases selection by users allowing them to drill down a particular cluster formed in the projection (Poco et al., 2012).

Although, for abstract data sets, 2D or point views are more common, the use of the third dimension (or third coordinate plane) can and most likely should be considered to allow additional representational power to the user.

In that sense, landscape views, besides being a common representation for actual terrain and geographical applications, can provide support to interpretation of various types of information. Some of the advantages are:

- supports perception of distribution according to a selected concept or variable;

- allows global description of similarity data mapped to 2D planes;

- lends itself to summarization of data sets;

- through artifacts describing regions, levels and distributions, potentially generates additional insights during exploration.

Current 2D mappings based on multidimensional projections have evolved greatly in recent years, and it is our argument that the new capabilities of these recent approaches improve the potential for coherent and informative landscape views. The reason is twofold:

- recent projections are more precise, forming groups of similar objects with less disturbance of neighborhoods, which eases exploration, and

- projections are nowadays also faster, making landscape views an alternative analysis tools for multilevel interaction.

In this article we illustrate the pipeline for creating and presenting landscape views that take advantage of the descriptive nature of multidimensional projections and demonstrate the artifacts that support better landscape exploration for applications such as interpretation of text and image collections.

3   Enhanced landscape construction

Starting from a set of data objects, we build a landscape through the pipeline shown in Figure 1. Each data object is processed to obtain a vector, that typically has many dimensions. Then a projection technique is used to obtain a 2D space, which is clustered and segmented. An additional data dimension is used to generate a landscape for display. If data objects are images or texts, then a word cloud or an image mosaic may be used as landscape texture. We discuss each step of the pipeline bellow.

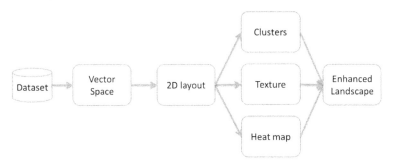

Figure 1: Enhanced landscape construction pipeline.

Data and vector space   Obtaining a vector from each data object typically involves extracting features and evaluating some measures. For text, term frequency-inverse document frequency (TF-IDF) is often used (Salton & Buckley, 1988). For images, descriptors like SIFT (Li & Wang, 2003) and BIC (Stehling et al., 2002), as well as many others, may be applied to obtain descriptive vectors for objects. The resulting vector space is usually a high-dimensional space.

For instance, a dataset with extracts of scientific articles, containing title, authors, abstract and references, from journals devoted to Case-based Reasoning (CBR), Inductive Logic Programming (ILP), Sonification (SON) and Information Retrieval (IR) was first processed for stop-words removal, stemming, Luhn's cut of terms with highest and lowest frequencies (Luhn, 1958) and then TF-IDF was evaluated. The resulting vector-space has 675 vectors with 1,423 dimensions. We will refer to this dataset as CBR+. The class of each article reflects origin of the document, then, in terms of subjects, some articles may not be in the most suitable class or should belong to more than one class. It is often the case in practice that pre-classification, if one exists, should not be taken strictly.

2D layout – mapping higher dimensions to visual 2D spaces   The next step is to project the vector-space, obtaining a 2D space. This is the role of a multidimensional projection, which strives to maintain certain properties of the original space in the projected space. For instance, Figure 2 shows different projections for the CBR+ dataset. The depicted layouts were produced by LSP, Principal Component Analysis (PCA – Jolliffe, 2002) and Self-Organizing Maps (SOM – Kohonen et al., 2001).

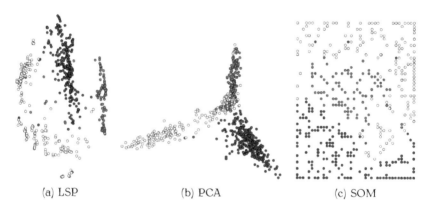

(a) LSP   (b) PCA   (c) SOM

Figure 2: Projections of CBR+. Color indicates pre-classification: CBR is dark-blue, ILP if light-blue, IR is yellow and SON is red.

Clustering – finding susitable groups of objects in projections   Clustering is the next step in the process. The goal is to define groups that allow partitioning the space into regions that will enhance data organization for exploration and navigation in data space.

Although any clustering algorithm may be used, a clustering algorithm that favors a smooth partitioning of the space into regions is more suitable. We propose two algorithms based on a Voronoi diagram. One defines clusters automatically and the other takes user guidance for defining the clusters. Both are very efficient computationally because a Voronoi diagram may be built in $O(n \lg n)$ time and the resulting structure is a planar map.

The first algorithm, called Voronoi Clustering, is greedy and distance-based. The algorithm starts with a Voronoi diagram where each point is a site (thus each diagram's region has exactly one data point). The algorithm visits every point p in arbitrary order and considers its neighbors. If the distance between p and a neighbor q is smaller than an input threshold then the regions of p and q are united.

The second algorithm, called Projection Seed Clustering, relies on the user to provide cluster seeds, visually selecting points on the projection display. Each cluster seed will give rise to a cluster. Suppose a graph G where each site is a vertex and such that there is a weighted edge between two vertices if their region in the Voronoi diagram is adjacent. Edge weights are distances. For each non-seed point p, evaluate the shortest-path from p to each seed s, and record the edge e_s in the shortest-path having the largest weight. Then add p to the region of the seed whose path from p contains the edge with smallest weight among those recorded for the shortest-paths, breaking ties arbitrarily. Although this algorithm resembles a shortest-path evaluation in a graph, the graph must not be built explicitly; using the Voronoi diagram and a queue leads to a very efficient implementation. Figure 3 shows the Voronoi and seed clustering for CBR+ applied after LSP.

When the input is a pre-classified dataset, it may be an option to skip clustering altogether. After projection, the 2D space may be segmented to produce a partition of the display, as those obtained with the clustering algorithms described above. The following algorithm will produce a segmentation of the 2D space from labeled (colored) points, with the premise that distance in the layout takes precedence over class information. Starting with a Voronoi diagram where each point is a site, unite adjacent regions whose site belong to the same class. Then, while either there are regions with too few points or there is a large number of regions, unite the smallest region with the closest one. An example output is shown in Figure 4.

There are other scenarios for using the above segmentation algorithm.

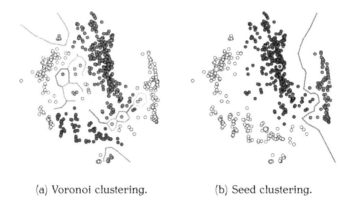

(a) Voronoi clustering.   (b) Seed clustering.

Figure 3: Voronoi and Seed clustering of CBR+ after LSP (from Figure 2a. The Voronoi clustering distance parameter value is 5% of the largest distance between any two points. In the seed clustering display, seeds are highlighted. Colors indicate formed clusters.

One of them is to reflect, in 2D space, clusters that approximate a cluster in the original space. Points are clustered in original space, which may give a precision gain, projected in 2D, and then subject to this segmentation algorithm, taking each cluster as a class, to obtain a more regular partitioning of the space into regions. Another scenario is using this segmentation to reduce the number of clusters obtained with Voronoi clustering or any other clustering algorithm.

The clustering algorithms described above may be applied recursively to refine the layout in as many levels as wanted, limited by the number of levels or the number of points in each region. For instance, Figure 5 shows a diagram partitioned in three levels.

The multi-level clustering after segmentation generates as a strategy to create multi-level presentations, supporting the generation of landscape layouts in larger scale.

Landscaping The layout may be enriched by mapping point color or attribute to height, thus generating a surface whose shape is indicative of data relations. We have considered two different heat maps that help the exploration of the layout. One is general applicable: a density heat map. The other, for text, is a term-frequency heat map. Both are constructed evaluating a scalar for each point and mapping such scalars onto a color scale. For instance, in Figure 6 the CBR+ dataset is shown under different

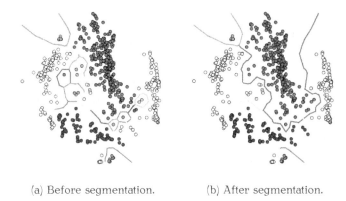

(a) Before segmentation.   (b) After segmentation.

Figure 4: The application of segmentation on a previously clustered space. The number of clusters was reduced by grouping small clusters to the closer ones.

Figure 5: A region partitioned into three levels of clustering. Clusters are divided by a thick continuous curve in the first level, by a blue discontinuous curve in the second level and by a thin discontinuous magenta curve in the third level.

heat maps. The scalar for density was built using the Epanechnikov kernel function (Silverman, 1986). The scalar for term-frequency was evaluated as the number of occurrences of one or more words (user input) in a document divided by the number of words in the document.

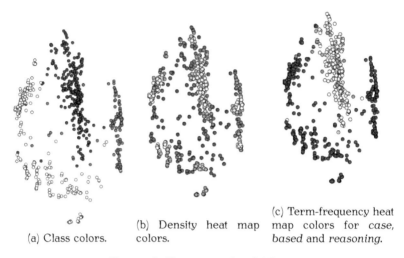

(a) Class colors.

(b) Density heat map colors.

(c) Term-frequency heat map colors for *case*, *based* and *reasoning*.

Figure 6: Heat maps for CBR+.

The landscape is built from the 2D space, mapping one attribute to the third dimension. For instance, Figure 7 shows a view of a landscape built for CBR+ mapping density to height. The cursor is positioned on a point, whose neighbors are highlighted by connecting lines and which label is shown. This type of cursor has been known as spider cursor (Minghim et al., 2005). Other attributes may be used to define height as well, including search terms for texts. This layout allows a global analysis relating the data set to an attribute (e.g. group density or term) as well as individual analysis of both groups and points.

We also propose combining the landscape with a texture that summarizes regions and adds more information to the layout, enhancing navigation and exploration. For text datasets, word clouds will provide a perfect texture for exploration. Figure 8 shows a view of the CBR+ density landscape enhanced by a word cloud built with the algorithm proposed by Paulovich et al. (2012), which generates word clouds in planar regions. Rotating the view allows alternating between a clearer view of the landscape slopes and valleys and a clearer view of words in each region. The height factor can also be removed when the association with the mapped attribute is no longer of interest.

The bordering lines of regions defined by clusters and segmentation also improve navigation and help preserving the mental model. Borders may enhance the perception of regions of interest and help preserving

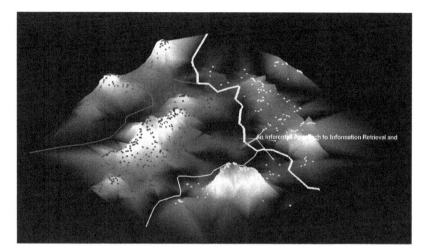

Figure 7: Landscape for CBR+, with density mapped to height.

reference points as the layout is rotated and zoomed, alleviating the effects of occlusion and flatness imposed by angled perspectives. Groups can be selected at once via the boundaries of the regions detected in the current segmentation. In the figures those boundaries are shown as linear curves.

Figure 8: Landscape with word cloud texture and clusters for CBR+, with density mapped to height.

Another example appears in Figure 9, that shows a landscape for a news dataset consisting of 1,771 RSS news feeds collected from BBC, CNN, Reuters and Associated Press during June and July of 2011. After processing, each document was turned into a vector of 1,084 dimensions.

A natural texture for image datasets is an image mosaic, such as those proposed by Tan et al. (2012). Figure 10 illustrates the concept for images in the Corel dataset, that has 1,000 images divided in 10 classes (African tribes, beaches, buildings, buses, dinosaurs, elephants, flowers horses, mountains and food), from which 150 dimensions were extracted through SIFT descriptors.

A summary may also be provided for a set of videos resulting also in a mosaic of images extracted from the video by summarizing algorithms. Music or other sound recordings may be summarized by a word cloud of tags, with added aural summaries played when a cursor is positioned on regions or points.

In general, a dataset may be summarized in many ways. Numeric attributes may be mapped on the landscape themselves or displayed as dispersion curves or histograms, for instance. Categorical attributes may be displayed as tag clouds as well.

Naturally, on the top of the landscape displays that we propose here, a series of symbols and glyphs may be added to enhance some characters of each dataset. While the effectiveness of adding more elements to the display must be considered for each dataset, particular applications may benefit from additional information on the display.

Finally, the landscape display, coupled with summarization and segmentation algorithms, may open the possibility for multi-level presentations. For instance, summarization can occur in various levels. While at the top level a general view, as well as the large parts of the data set, are displayed, the user may choose a region to focus his or her exploration. Figure 11 shows a top view of the news data set presented before. In that figure, the top level is presented and subjects of the various regions are displayed via a co-occurrence based topic extraction algorithm. A group is then selected, and for that group, the display is repeated, with an added level of detail given by new clustering followed by segmentation. The strategy is prone to be applied for multi-level point-based and surface-based displays.

4 Conclusion

Landscapes for information visualization have been encouraged by many authors, although without support from experiments with user subjects.

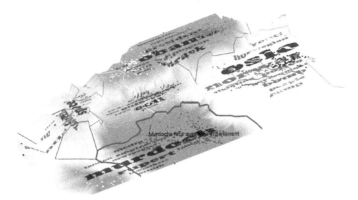

(a) News landscape with word clouds, and density as height.

(b) Flatter version of the news landscape.

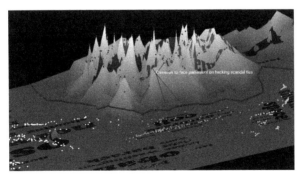

(c) News landscape with search terms frequencies (murdoch and hacking) mapped to height.

Figure 9: Landscapes from RSS news feeds.

(a) Mosaic over projection. Central picture of the cursor is shown.

(b) Mosaic with density landscaping. Central picture of the cursor is shown.

(c) Mosaic with display of all pictures touched by the spider cursor.

Figure 10: Landscape with mosaic texture and clusters for the Corel dataset, a 10-class image data set. Point colors indicate classes.

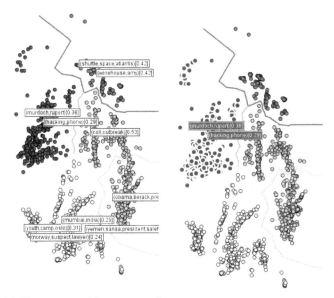

(a) News map with groups of news subjects, identified by topics.

(b) Selection of a group of news feeds.

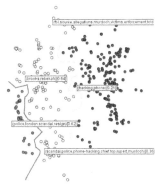

(c) New level of grouping in the selected set, after clustering and segmentation.

Figure 11: Drilling down on the map levels.

We believe that the newest projection techniques that have been developed recently renew the strength of mapping data through the concept of landscapes, as they are both fast and precise enough to enable real-time construction of layouts recursively and in a very large scale. Visual guidance, as those we introduced in this article, namely the division of the landscape in regions and the use of texture, also help in avoiding disorientation in 3D and increases contextual overviews. The view lends itself to multi-level visualizations, allowing for future development of visualization in larger scale.

References

Bischoff, U., Diakopoulos, N., Loesch, F. & Zhou, Y. (2004), ThemeExplorer: A tool for understanding the history of the field of information visualization, *in* 'Proc. InfoVisFun'.

Boyack, K. W., Wylie, B. N. & Davidson, G. S. (2002), 'Domain visualization using VxInsight for science and technology management', *Journal of the American Society for Information Science and Technology* **53**(9), 764–774.

Chalmers, M. (1993), Using a landscape metaphor to represent a corpus of documents, *in* 'Proc. European Conference on Spatial Information Theory', pp. 377–390.

Cockburn, A. & McKenzie, B. (2001), 3D or not 3D?: Evaluating the effect of the third dimension in a document management system, *in* 'Proc. SIGCHI Conference on Human Factors in Computing Systems', pp. 434–441.

Davidson, G. S., Hendrickson, B., Johnson, D. K., Meyers, C. E. & Wylie, B. N. (1998), 'Knowledge mining with vxinsight: Discovery through interaction', *Journal of Intelligent Information Systems* **11**, 259–285.

Fabrikant, S. I., Maggi, S. & Montello, D. R. (2014), 3D network spatialization: Does it add depth to 2D representations of semantic proximity?, *in* 'Proc. GIScience', pp. 34–47.

Fabrikant, S. I., Montello, D. R. & Mark, D. M. (2010), 'The natural landscape metaphor in information visualization: The role of commonsense geomorphology', *Journal of the American Society for Information Science and Technology* **61**(2), 253–270.

Fruchterman, T. M. J. & Reingold, E. M. (1991), 'Graph drawing by force-directed placement', *Software - Practice and Experience* **21**(11), 1129–1164.

Fung, D. C., Hong, S.-H., Koschutzki, D., Schreiber, F. & Xu, K. (2008), '2.5d visualisation of overlapping biological networks', *Journal of Integrative Bioinformatics*.

Hall, M. & Clough, P. (2013), Exploring large digital library collections using a map-based visualisation, *in* 'Proc. International Conference on Theory and Practice of Digital Libraries', pp. 216–227.

Havre, S., Hetzler, E., Whitney, P. & Nowell, L. (2002), 'Themeriver: visualizing thematic changes in large document collections', *IEEE Transactions on Visualization and Computer Graphics* **8**(1), 9–20.

Jaffe, A., Naaman, M., Tassa, T. & Davis, M. (2006), Generating summaries and visualization for large collections of geo-referenced photographs, *in* 'Proc. ACM International Workshop on Multimedia Information Retrieval', pp. 89–98.

Jianu, R. & Laidlaw, D. H. (2013), 'What google maps can do for biomedical data dissemination: examples and a design study', *BMC Research Notes* **6**(1), 1–14.

Jolliffe, I. T. (2002), *Principal Component Analysis*, 2 ed., Springer-Verlag.

Kohonen, T., Schroeder, M. R. & Huang, T. S., eds (2001), *Self-Organizing Maps*, 3rd ed., Springer-Verlag.

Li, J. & Wang, J. Z. (2003), 'Automatic linguistic indexing of pictures by a statistical modelin approach', *IEEE Transactions on Pattern Analysis and Machine Intelligence* **25**, 1075–1088.

Luhn, H. P. (1958), 'The automatic creation of literature abstracts', *IBM Journal of Research and Development* **2**(2), 159–165.

Lynch, K. (1960), *The Image of the City*, MIT Press.

Mazza, R. (2009), *Introduction to Information Visualization*, Springer-Verlag.

Minghim, R., Levkowitz, H., Nonato, L. G., Watanabe, L., Salvador, V., Lopes, H., Pesco, S. & Tavares, G. (2005), Spider cursor: A simple verstile interaction tool for data visualization and exploration, *in* 'Proceedings of the 3rd International Conference on Computer Graphics and Interaction Techniques in Australasia and Southeast Asia - GRAPHITE', ACM Press, Dunedin, New Zeland, pp. 307–314.

Nam, J. E. & Mueller, K. (2013), 'Tripadvisor N-D: A tourism-inspired high-dimensional space exploration framework with overview and detail', *IEEE Transactions on Visualization And Computer Graphics* 19(2), 291–305.

Pampalk, E. (2001), Islands of Music Analysis, Organization, and Visualization of Music Archives, PhD thesis, Technischen Universität Wien.

Paulovich, F. V. & Minghim, R. (2006), Text map explorer: a tool to create and explore document maps, *in* 'Proc. IEEE Symposium on Information Visualization', pp. 245–251.

Paulovich, F. V., Nonato, L. G., Minghim, R. & Levkowitz, H. (2006), Visual mapping of text collections through a fast high precision projection technique, *in* 'Proc. IEEE Symposium on Information Visualization', pp. 282–290.

Paulovich, F. V., Telles, G. P., Toledo, F. M. B., Minghim, R. & Nonato, L. G. (2012), 'Semantic wordification of document collections', *Computer Graphics Forum* **31**, 1145–1153.

Piringer, H., Kosara, R. & Hauser, H. (2004), Interactive focus+context visualization with linked 2D/3D scatterplots, *in* 'Proc. International Conference on Coordinated and Multiple Views in Exploratory Visualization', pp. 49–60.

Poco, J., Eler, D. M., Paulovich, F. V. & Minghim, R. (2012), 'Employing 2D projections for fast visual exploration of large fiber tracking data', *Computer Graphics Forum* **31**(3), 1075–1084.

Risden, K., Czerwinski, M. P., Munzner, T. & Cook, D. B. (2000), 'An initial examination of ease of use for 2D and 3D information visualizations of web content', *International Journal of Human-Computer Studies* **53**(5), 695–714.

Robertson, G., Czerwinski, M., Larson, K., Robbins, D. C., Thiel, D. & van Dantzich, M. (1998), Data mountain: Using spatial memory for document management, *in* 'Proc. UIST'.

Salton, G. & Buckley, C. (1988), 'Term-weighting approaches in automatic text retrieval', *Information Processing & Management* **24**(5), 513–523.

Silverman, B. W. (1986), *Density Estimation for Statistics and Data Analysis*, Chapman & Hall.

Skupin, A. (2004), 'The world of geography: Visualizing a knowledge domain with cartographic means', *PNAS* **101**(1), 5274–5278.

Skupin, A. & Buttenfield, B. P. (1996), Spatial metaphors for visualizing very large data archives, *in* 'Proc. GIS-LIS International Conference', pp. 607–617.

Stehling, R. O., Nascimento, M. A. & Falcão, A. X. (2002), A compact and efficient image retrieval approach based on border/interior pixel classification, *in* 'Proc. International Conference on Information and Knowledge Management', pp. 102–109.

Tan, L., Song, Y., Liu, S. & Xie, L. (2012), 'Imagehive: Interactive content-aware image summarization', *IEEE Computer Graphics and Applications* **32**(1), 46–55.

Tory, M., Kirkpatrick, A. E., Atkins, M. S. & Moller, T. (2006), 'Visualization task performance with 2D, 3D, and combination displays', *IEEE Transactions on Visualization and Computer Graphics* **12**(1), 2–13.

Tory, M., Sprague, D. W., Wu, F., So, W. Y. & Munzner, T. (2007), 'Spatialization design: Comparing points and landscapes', *IEEE Transactions on Visualization and Computer Graphics* **13**(6), 1262–1269.

Tory, M., Swindells, C. & Dreezer, R. (2009), 'Comparing dot and landscape spatializations for visual memory differences', *IEEE Transactions on Visualization and Computer Graphics* **15**(6), 1033–1040.

Westerman, S. J. & Cribbin, T. (2000), 'Mapping semantic information in virtual space: Dimensions, variance and individual differences', *International Journal of Human-Computer Studies* **53**(5), 765–787.

Wise, J. A., Thomas, J. J., Pennock, K., Lantrip, D., Pottier, M., Schur, A. & Crow, V. (1995), Visualizing the non-visual: Spatial analysis and interaction with information from text documents, *in* 'Proc. IEEE Symposium Information Visualization', pp. 51–58.

A study on edge detection algorithms modified by non-standard neighbourhood configurations for use in SAR imagery[‡]

Sandra Aparecida Sandri *   Gilberto P. Silva Junior *

* Laboratório Associado de Computação e Matemática Aplicada, Instituto Nacional de Pesquisas Espaciais, Avenida dos Astronautas, N. 1.758, CEP:12227 − 010, São José dos Campos, São Paulo, Brasil
sandra.sandri@inpe.br, gp7junior@gmail.com

1 Introduction

SAR images (Lee & Pottier, 2009) have been increasingly used in remote sensing applications. This is due, among other reasons, to the fact that SAR sensors, in contrast to optical ones, are not so adversely affected by atmospheric conditions and the presence of clouds, allowing them to be used at any time of day or night. SAR systems generate the image of a target area by moving along a trajectory, usually linear, while transmitting and receiving pulses in lateral looks towards the ground, in either horizontal (H) or vertical (V) polarizations (Richards, 2009). PolSAR systems (*Polarimetric Synthetic Aperture Radar*) generate images in HH, VV, HV and VH polarizations (instead of only HH and VV as in previous technology), although in many applications HV and VH are considered to be the same.

Due to the inherent characteristics of radar technology, SAR images contain *speckle* (Mott, 2006), a multiplicative non-Gaussian noise that is proportional to the intensity of the received signal. Speckle degrades the visual quality of the displayed image by sudden variations in image intensity. It can be reduced with multiple looks in the generation of complex images, causing degradation in spatial resolution. Speckle can also be reduced with the use of filters. The most well-known filtering method for SAR images is the so-called Lee (or sigma) filter, introduced in (Lee, 1983); modified versions of it, such as the "Enhanced Lee" filter, proposed

[‡]The authors are very grateful to Humberto Bustince, Wagner Barreto Silva, Leonardo Torres and Corina Freitas for help in the preparation of this manuscript. The authors are also thankful to CAPES, CNPq and FAPESP for financial support.

by Lopes et al. (1990), very often provide even better results. A recently proposed filter (Torres et al., 2014) has a nonlocal means approach for PolSAR image speckle reduction based on stochastic distances.

Edge detection is an important area of image analysis, aiming at automatically identifying sharp differences in the information associated with adjacent pixels in an image (Gonzalez & Woods, 2006). Many edge detection methods have been created for optical imagery. The most important one is the so-called Canny method (Canny, 1986). Recent edge detection methods include those stemming from Computational Intelligence, such as those inspired on Newton's Universal Law of Gravity. Sun et al. (2007) proposed the gravitational edge detection method, which was then extended by Lopez-Molina et al. (2010) using operators from the Fuzzy Sets Theory literature. Other methods based on fuzzy sets are due to Danková et al. (2011), with a fuzzy based function, called the F-transform, and to Barrenechea et al. (2011) who propose the use of interval-valued fuzzy relations. For SAR imagery, however, there exist few edge detection algorithms specifically built to handle them (Fu et al., 2012).

Recently (Silva Junior et al., 2014a,b), we have studied the influence of filtering SAR images before the use of the edge detection methods due to Lopez-Molina et al and Canny, mentioned previously. Experiments were carried on both the original images and the ones obtained by aggregating the images from HH, VV and HV polarizations, before and after edge detection. In those works, Lopez-Molina method was also addressed when modified by computing the pixel values in a 3×3 target window as the average of the pixels in the 9×9 window from Fu neighbourhood configuration (Fu et al., 2012). Both the Enhanced Lee and Torres filters were employed, with a slight advantage for the former.

The present work is an extension of (Silva Junior et al., 2014a,b), addressing the modification of Lopez Molina gravitational edge detection method by the non-standard neighbourhood configurations of (Fu et al., 2012), (Dimou et al., 2000) and (Nagao & Matsuyama, 1979). We present experiments on a synthetic mosaic obtained from a real scene, considering both filtered and unfiltered images. In the experiments, the modification of the gravitational algorithm, using any of the three neighbourhood configurations, yielded better results, according to the Baddeley Delta metric (BDM – Baddeley, 1992), than those obtained with no modification, which shows that the approach is promising.

2   General formalization of edge detection algorithms modified by non-standard neighbourhood configurations

In the present work, we address the following modification of an edge detection algorithm: given a $k \times k$ window in an image, the values considered for use in the algorithm for the k^2 positions in the window are no longer the ones from those positions in the image, but the mean values of the k^2 regions taken from a larger window in that image, according to a given neighbourhood configuration. Below, we give a general formalization of the modification of an edge detection algorithm with a non-standard neighbourhood configuration.

Let p_0 denote the pixel at the center of a $k \times k$ window in a given image I. Let N_0 be a given $n \times n$ neighbourhood configuration around p_0 in I, such that $n \geq k$. Let P_i denote the set of pixels in the i-th region in neighbourhood configuration N_0. Let $v(.)$ denote the value associated to a given pixel in image I. In the modified version of a given edge detection algorithm, each pixel p_i in the $k \times k$ window becomes associated to $\mu(i)$, the average value of the pixels in P_i, defined as

$$\mu(i) = \frac{1}{\mid P_i \mid} \times \sum_{p_j \in P_i} v(p_j), 0 \leq i \leq k^2 - 1 \qquad (1)$$

Figure 1 brings the resulting working window after the application of the modification above for a 3×3 window in a given image. Figures 2, 3 and 4, bring the non-standard neighbourhood configurations due to Fu et al. (2012), Dimou et al. (2000) and Nagao & Matsuyama (1979), respectively, with the regions labeled according to Figure 1. Note that for both Fu and Dimou neighbourhoods we have $\mid P_0 \mid = 1$, i.e. the values of the central pixels in the $k \times k$ and the $n \times n$ windows coincide, whereas for Nagao-Matsuyama neighbourhood we have $\mid P_0 \mid = 9$.

$\mu(1)$	$\mu(2)$	$\mu(3)$
$\mu(8)$	$\mu(p_0)$	$\mu(4)$
$\mu(7)$	$\mu(6)$	$\mu(5)$

Figure 1: Modified 3×3 neighbourhood (Silva Junior et al., 2014a)

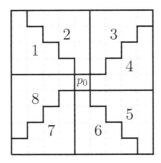

Figure 2: Fu's 9×9 neighbourhood configuration (Fu et al., 2012)

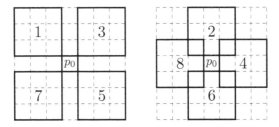

Figure 3: Dimou's 7×7 neighbourhood configuration (Dimou et al., 2000)

3   Materials and methods

3.1   Gravitational approach for edge detection

Newton's Universal Law of Gravity is described by Equation 2:

$$f_{1,2} = G \times \frac{m_1 \times m_2}{\|\vec{r}\|^2} \times \frac{\vec{r}}{\|\vec{r}\|}, \tag{2}$$

where m_1 and m_2 are the masses of two bodies, $\vec{r}$ is the vector connecting them, $\vec{f}_{1,2}$ is the gravitational force between them, $\|.\|$ denotes the magnitude of a vector, and G is the gravitational constant.

The gravitational edge detection approach is based on Newton's Universal Law of Gravity and was first proposed by Sun et al. (2007) in the context of optical images. In the analogy proposed by Sun et al. (2007), the bodies are the gray level values of pixels in a grid, G is a function of values of the pixels in a given window, the distance between any two adjacent pixels is equal to 1 and, when computing the resulting force of the pixel in the center of a window, the pixels outside that window are considered negligible.

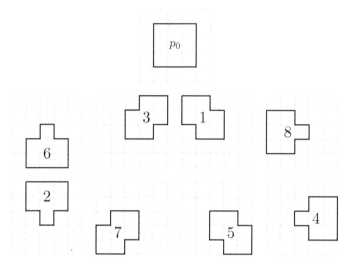

Figure 4: Nagao-Matsuyama's 5×5 neighbourhood conf. (Nagao & Matsuyama, 1979)

Lopez-Molina et al. (2010) extended Sun et al. (2007)'s approach, so that any Triangular Norm (operators that implement conjunction in Fuzzy Sets Theory (Dubois & Prade, 1988)) other than the product can be used in place of the product between the two masses, by first normalizing the gray level values to [0,1]. Another particularity from this approach is that it takes G as a normalization constant, calculated so as to guarantee that the forces resulting from always lie in [0,1]. Moreover, in the normalization of gray level values into [0,1], a small value δq is added beforehand on both the numerator and denominator, in this way avoiding pixels with value 0 to have too strong effect on neighbouring pixels. In the present work, we have adopted the particularities of the approach proposed by Lopes Molina et al, keeping however the T-norm product in Equation 2.

3.2 Quality assessment

Baddeley Delta metric (BDM – Baddeley, 1992) aims at measuring the dissimilarity of subsets of featured points. This metric presents advantages over accuracy, when used to compare two binary images, the one produced by an edge detection algorithm and the ideal one. Contrary to accuracy, BDM also considers borders close to the ideal ones in the computation, allowing for more flexibility.

Let $\mathbf{x}$ and $\mathbf{y}$ be two binary images, seen as mappings from Λ to $\{0, 1\}$, where Λ is a set of sites arranged in a grid (positions). Let ρ be a metric on Λ, such as the Euclidean distance, and $d(i, A)$ be the distance between a site i and a set $A \subseteq \Lambda$, defined as

$$d(i, A) = \min_{j \in A} \rho(i, j).$$

Let $b(\mathbf{x})$ denote the set of foreground sites in $\mathbf{x}$: $b(\mathbf{x}) = \{\mathbf{i} \in \Lambda \mid \mathbf{x_i} = 1\}$. BDM between $\mathbf{x}$ and $\mathbf{y}$, denoted as $\Delta_{p,w}$, is then defined as

$$\Delta_{p,w}(\mathbf{x}, \mathbf{y}) = \left(\frac{1}{|\Lambda|} \sum_{i \in \Lambda} |w(d(i, b(\mathbf{x}))) - w(d(i, b(\mathbf{y}))) |^p \right)^{\frac{1}{p}}, 1 \leq p \leq \infty \quad (3)$$

where w is a concave strictly increasing function satisfying $w(0) = 0$.

In the experiments presented here, pixels in the image boundaries are not taken into account in the BDM calculation and the parameters have been set as $w(t) = t$ and $p = 2$, like in (Lopez-Molina et al., 2010; Silva Junior et al., 2014a,b). Throughout the text, we display BDM results in [0,100] instead of [0,1], for readability sake.

3.3 Working image

In the experiments, we make use of a mosaic of synthetic images derived from a full polarimetric image in band L, taken from a scene of an agricultural area in the North of Brazil, depicting water and different types of vegetation, some of which in different stages of growth. Samples for each class were extracted from this image, and using these samples the parameters of the complex Wishart distribution were estimated for each class (da Silva et al., 2013).

Here we used 20 synthetic images simulated with the use of the parameters estimated for each class, the same ones used in (Silva Junior et al., 2014a,b). Figure 5 brings the RGB composition of the amplitude images A_{HH}, A_{HV} and A_{VV}, derived in this manner from polarizations HH, HV and VV.

4 Experimental results

In the following, we discuss the results obtained for the mosaic image, using the same notation as in (Silva Junior et al., 2014a):

- DB_{HH}, DB_{VV} and DB_{HV} denote the binary images obtained from images A_{HH}, A_{HV} and A_{VV};

Figure 5: Mosaic of synthetic amplitude images computed on polarizations HH, HV and VV (simulated using the equations in da Silva et al. (2013))

- ADB denotes the binary image obtained by first aggregating images A_{HH}, A_{HV} and A_{VV}, then applying an edge detection algorithm, followed by binarization;

- DAB denotes the binary image obtained by first applying an edge detection algorithm on images A_{HH}, A_{HV} and A_{VV}, then aggregating the results, followed by binarization.

Here we address the use of Lopez-Molina gravitational method with and without modifications with non-standard configurations. The output of Lopez-Molina method is an image with values in [0,1]. In order to obtain binary indicators of edges, the authors use a hysteresis transformation. Here, we use a simple threshold and search for the value in the [0.05, 0.15] interval which produces the best BDM.

We also use of filters proposed in (Torres et al., 2014) and (Lopes et al., 1990) in the experiments. They are applied as a preprocessing step, thus before aggregation for ADB and before detection for DB_{HH}, DB_{VV}, DB_{HV} and DAB. In both cases, the filters are applied on intensity values, which are then transformed in amplitude before further processing.

Table 1 brings the average BDM results for Lopez-Molina method, considering 20 simulated images, modified by Fu, Dimou and Nagao-Matsuyama neighbourhood configurations[1]. We can see that the modification of the gravitational algorithm using any of the three neighbourhood configurations improves the results obtained without the modification. In particular, the best results are obtained using Fu neighbourhood configuration. Moreover, we see that the joint use of the modified algorithms and filtering produces better results than their isolated use.

---

[1]The results without modification and those modified by Fu are taken from (Silva Junior et al., 2014a,b).

Table 1: Average BDM results for Lopez-Molina method without modification and modified by Fu, Dimou and Nagao-Matsuyama neighbourhood configurations; the standard deviation is inside parenthesis

	Original		
Image	No filter	Torres filter	Enh. Lee filter
DB_{HH}	33.89 (1.98)	26.61 (2.00)	38.97 (0.79)
DB_{VV}	32.35 (1.47)	28.95 (0.95)	43.65 (1.13)
DB_{HV}	31.95 (0.46)	27.14 (1.22)	32.26 (2.78)
DAB	29.26 (1.62)	25.91 (1.55)	27.71 (2.62)
ADB	31.50 (0.82)	26.63 (1.27)	18.24 (3.41)
	Fu		
Image	No filter	Torres filter	Enh. Lee filter
DB_{HH}	25.27 (0.76)	22.18 (0.48)	17.79 (3.05)
DB_{VV}	21.41 (1.97)	18.14 (0.77)	17.83 (2.54)
DB_{HV}	26.48 (1.00)	24.21 (0.65)	18.40 (5.75)
DAB	22.67 (2.29)	18.97 (1.62)	5.43 (1.68)
ADB	23.80 (2.23)	22.74 (0.50)	5.16 (0.36)
	Dimou		
Image	No filter	Torres filter	Enh. Lee filter
DB_{HH}	26.83 (1.57)	23.68 (0.76)	17.79 (3.52)
DB_{VV}	24.59 (1.45)	20.21 (1.98)	32.61 (1.47)
DB_{HV}	27.92 (0.90)	25.19 (1.10)	21.72 (4.85)
DAB	26.00 (1.48)	22.85 (1.70)	8.61 (1.80)
ADB	25.22 (0.63)	23.50 (0.59)	6.83 (2.64)
	Nagao-Matsuyama		
Image	No filter	Torres filter	Enh. Lee filter
DB_{HH}	28.06 (1.71)	24.80 (1.09)	25.89 (2.49)
DB_{VV}	28.26 (0.99)	24.27 (1.20)	37.56 (1.22)
DB_{HV}	31.48 (0.73)	27.60 (1.33)	24.43 (5.27)
DAB	27.54 (0.96)	25.53 (0.99)	13.43 (3.31)
ADB	27.14 (0.47)	24.79 (0.70)	13.73 (4.09)

Figures 6, 7, 8 and 9 respectively bring the negative images corresponding to the best results, according to BDM, obtained by the use of filters and the edge detection methods both with and without modification. We see that the best results by visual inspection coincide with the best results according to BDM.

The best binary image, according to BDM, depicted in Figure 7a, presents little noise and most of the regions are separated, even though the lines are rather thick. In the best results for the other configurations, the regions are not separated so well, but they produce thinner separation lines. When we compare the best results obtained from the edge detection algorithm without modification to those obtained with the modification, we see that the former produce not only the regions that are not very well separated but a larger amount of noise.

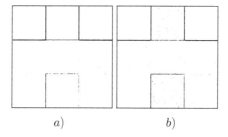

Figure 6: Best BDM results obtained with no modification: a) ADB with Enh. Lee filtering (BDM = 10.96), b) DAB with Enh. Lee filtering (BDM = 21.07)

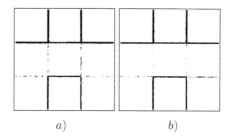

Figure 7: Best BDM results obtained using Fu neighbourhood configurations: a) ADB with Enh. Lee filtering (BDM = 3.05), b) DAB with Enh. Lee filtering (BDM = 3.00)

5   Conclusions

We addressed the issue of modifying the gravitational approach proposed by Lopez-Molina et al. (2010) with the non-standard neighbourhood configurations proposed by (Fu et al., 2012), (Dimou et al., 2000) and (Nagao & Matsuyama, 1979), for decreasing noise in polarimetric Synthetic Aperture Radar – SAR imagery. We also studied the effect of filtering the images

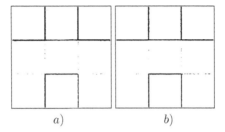

Figure 8: Best BDM results obtained Dimou neighbourhood configurations: a) ADB with Enh. Lee filtering (BDM = 3.48), b) DAB with Enh. Lee filtering (BDM = 3.92)

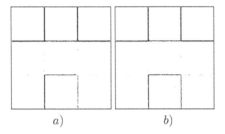

Figure 9: Best BDM results obtained Nagao-Matsuyam neighbourhood configurations, a) DAB with Enh. Lee filtering (BDM = 6.50), b) ADB with Enh. Lee filtering (BDM = 8.17)

prior to edge detection by two filters: Enhanced Lee (Lopes et al., 1990) and Torres (Torres et al., 2014). The methods were applied on 20 simulations of a synthetic image, derived from (da Silva et al., 2013). Using both visual inspection and the Baddeley Delta metric, we verify that the combination with the Lopez-Molina technique with the 9×9 neighbourhood proposed by Fu et al. (2012) and preprocessing with the Enhanced Lee filter produces the best results but that modifying the algorithm with any of the three non-standard neighbourhood configurations produces better results than the use of the algorithm, without modification. Our main conclusion is that the joint use of the modified algorithms and filtering produces better results than their isolated use.

Future work includes the use of other means for comparing the results, such as the one proposed recently by Buemi et al. (2014). We also intend to better address the issue of aggregation. Here we have dealt exclusively with the aggregation of non-binary images, using the arithmetic means in strategies ADB and DBA proposed in (Silva Junior et al., 2014*a*). In the future, we intend to explore aggregation considering families of operators in general, such as weighted means, ordered weighted means (OWA), T-norms and T-conorms (Dubois & Prade, 1988), and also strategy DBA, in which we aggregate the edge images, obtained after edge detection and binarization.

This work is an extension of (Silva Junior et al., 2014*a,b*), and is part of an ongoing project that aims at investigating the use of edge detection methods derived from Computational Intelligence techniques for use of SAR images. In the future, we intend to verify the usefulness of other techniques for edge detection in radar imagery, such as the one proposed by Lopez-Molina et al. (2011), involving the use of fuzzy sets to produce image gradients.

References

Baddeley, A. J. (1992), An error metric for binary images, *in* W. Förstner & H. Ruwiedel, eds, 'Robust Computer Vision: Quality of Vision Algorithms', Wichmann, Karlsruhe, pp. 59–78.

Barrenechea, E., Bustince, H., Baets, B. D. & Lopez-Molina, C. (2011), 'Construction of interval-valued fuzzy relations with application to the generation of fuzzy edge images', *IEEE Transactions on Fuzzy Systems* **19**(5), 819–830.

Buemi, M. E., Frery, A. C. & Ramos, H. S. (2014), 'Speckle reduction with adaptive stack filters', *Pattern Recognition Letters* **36**, 281–287. DOI 10.1016/j.patrec.2013.06.005. ISSN 0167-8655. URL http://dx.doi.org/10.1016/j.patrec.2013.06.005.

Canny, J. (1986), 'A computational approach to edge detection', *IEEE Transactions on Pattern Analysis and Machine Intelligence* **8**(6), 679–698. DOI 10.1109/TPAMI.1986.4767851.

da Silva, W. B., da Costa Freitas, C., Sant'Anna, S. J. S. & Frery, A. C. (2013), 'Classification of segments in PolSAR imagery by minimum stochastic distances between Wishart distributions', *IEEE Journal of Selected Topics in Applied Earth Observations and Remote Sensing* **6**(3), 1263–1273.

Danková, M., Hodáková, P., Perfilieva, I. & Vajgl, M. (2011), Edge detection using F-transform, *in* '2011 11th International Conference on Intelligent Systems Design and Applications (ISDA)', IEEE, pp. 672–677.

Dimou, A., Uzunoglou, N., Frangos, P., Jager, G. & Benz, U. (2000), Linear features' detection in sar images using fuzzy edge detector (fed), *in* 'Space-Based Observation Technology'.

Dubois, D. & Prade, H. (1988), *Possibility Theory: An Approach to Computerized Processing of Uncertainty*, Plenum Press, New York, USA.

Fu, X., You, H. & Fu, K. (2012), 'A statistical approach to detect edges in SAR images based on square successive difference of averages', *IEEE Geoscience and Remote Sensing Letters* 9(6), 1094–1098. DOI 10.1109/LGRS.2012.2190378. ISSN 1545-598X.

Gonzalez, R. C. & Woods, R. E. (2006), *Digital Image Processing (3rd Ed.)*, Prentice-Hall, Inc., Upper Saddle River, NJ, USA.

Lee, J. & Pottier, E. (2009), *Polarimetric Radar Imaging: From Basics to Applications*, Optical Science and Engineering, Taylor & Francis. URL http://books.google.com.br/books?id=1nAvp2HW_gwC.

Lee, J.-S. (1983), 'A simple speckle smoothing algorithm for synthetic aperture radar images', *IEEE Transactions on Systems, Man and Cybernetics* 13(1), 85–89. DOI 10.1109/TSMC.1983.6313036. ISSN 0018-9472.

Lopes, A., Touzi, R. & Nezry, E. (1990), 'Adaptive speckle filters and scene heterogeneity', *IEEE Transactions on Geoscience and Remote Sensing* 28(6), 992–1000.

Lopez-Molina, C., Baets, B. D. & Bustince, H. (2011), 'Generating fuzzy edge images from gradient magnitudes', *Computer Vision and Image Understanding* 115(11), 1571–1580. DOI http://dx.doi.org/10.1016/j.cviu.2011.07.003. ISSN 1077-3142.

Lopez-Molina, C., Bustince, H., Fernandez, J., Couto, P. & Baets, B. D. (2010), 'A gravitational approach to edge detection based on triangular norms', *Pattern Recognition* 43(11), 3730–3741. DOI http://dx.doi.org/10.1016/j.patcog.2010.05.035. ISSN 0031-3203.

Mott, H. (2006), *Remote Sensing with Polarimetric Radar*, John Wiley & Sons.

Nagao, M. & Matsuyama, T. (1979), 'Edge preserving smoothing', *Computer Graphics and Image Processing* **9**(4), 394–407. ISSN 0146-664X.

Richards, J. (2009), *Remote Sensing with Imaging Radar*, Signals and Communication Technology, Springer.

Silva Junior, G. P., Frery, A. & Sandri, S. (2014a), 'A study on the use of optical images based edge detection techniques on synthetic aperture radar images', *submitted*.

Silva Junior, G. P., Frery, A. & Sandri, S. (2014b), Synthetic aperture radar edge detection with Canny's procedure and a gravitational approach, *in* '11th International FLINS Conference on Decision Making and Soft Computing (FLINS)', pp. 149–154. DOI 10.1142/9789814619998_0027.

Sun, G., Liu, Q., Ji, C. & Li, X. (2007), 'A novel approach for edge detection based on the theory of universal gravity theory of universal gravity', *Pattern Recognition* **40**(10), 2766–2775.

Torres, L., Sant'Anna, S. J. S., Freitas, C. C. & Frery, A. C. (2014), 'Speckle reduction in polarimetric SAR imagery with stochastic distances and nonlocal means', *Pattern Recognition* **47**, 141–157. DOI 10.1016/j.patcog.2013.04.001.

www.ingramcontent.com/pod-product-compliance
Lightning Source LLC
LaVergne TN
LVHW012330060326
832902LV00011B/1809